Molecular Structure, Function, and Assembly of the ATP Synthases

Molecular Structure, Function, and Assembly of the ATP Synthases

Edited by
Sangkot Marzuki
Monash University
Clayton, Victoria, Australia

Plenum Press • New York and London

Library of Congress Cataloging-in-Publication Data

International Seminar/Workshop on the Molecular Structure, Function,
  and Assembly of ATP Synthases (1987 : Honolulu, Hawaii)
    Molecular structure, function, and assembly of the ATP synthases /
edited by Sangkot Marzuki.
      p.   cm.
    "Proceedings of the International Seminar/Workshop on the
Molecular Structure, Function, and Assembly of ATP Synthases, held
April 22-24, 1987, in Honolulu, Hawaii"--T.p. verso.
    Includes bibliographical references.
    ISBN 0-306-43499-7
    1. Adenosine triphosphatase--Congresses. 2. Adenosine
triphosphatase genes--Expression--Congresses.   I. Marzuki, Sangkot.
II. Title.
QP609.A3I574 1967
574.87'328--dc20   90-6709
                                                        CIP

Proceedings of the International Seminar/Workshop on the
Molecular Structure, Function, and Assembly of ATP Synthases,
held April 22-24, 1987, in Honolulu, Hawaii

© 1989 Plenum Press, New York
A Division of Plenum Publishing Corporation
233 Spring Street, New York, N.Y. 10013

Printed in the United States of America

PREFACE

In recent years, the ATP synthase (H^+-ATPase, F_oF_1-ATPase) has been the subject of intensive investigations in many laboratories. The major reason for this lies in the fact that this enzyme complex catalyses one of the most important reactions in living cells, namely the synthesis of ATP utilizing the energy from an electrochemical transmembrane H^+ gradient, generated by the cellular respiratory chain or by the light reactions of photosynthetic organisms. The mechanism by which the H^+ motive force is utilized to drive the synthesis of ATP is one of the major unsolved problems in biochemistry. Thus, the fundamental information concerning the molecular structure and the mechanism of assembly of the ATP synthase is of major significance in cell biology.

A seminar/workshop on the Molecular Structure, Function and Assembly of the ATP synthases was held in April, 1987 at the East-West Center, University of Hawaii, Honolulu, Hawaii, to promote exchange of information between laboratories actively engaged in the study of the ATP synthases, and to provide a forum for discussion and coordination of data derived from molecular, genetic and biochemical approaches used in different laboratories. This volume summarizes the result of the seminar/workshop, in the form of a collection of papers presented at the meeting, and provides an overview of current work in this rapidly progressing area of research.

The seminar/workshop was organized with the main sponsorships of the Department of Science (Australia), the National Science Foundation (USA) and the Japanese Society for the Promotion of Science, obtained under the US/Australia Agreement of Scientific and Technical Cooperation and the US/Japan Science Program. The sponsorship of the International Union of Biochemistry (IUB cosponsored symposium No. 164) is gratefully acknowledged. I wish to thank Anthony W. Linnane, D. Rao Sanadi, Yasuo Kagawa, Richard J. Guillory and Masamitsu Futai for their efforts in the organization of the seminar/workshop.

S.M.

CONTENTS

MOLECULAR GENETICS OF THE ATP SYNTHASE

POLYPEPTIDE SYNTHESIS, IMPORT AND ASSEMBLY

STRUCTURE AND FUNCTION OF ATP SYNTHASE

THE F$_o$ SECTOR AND PROTON MOTIVE COUPLING

ATP SYNTHASE REACTION MECHANISMS

MOLECULAR GENETICS OF THE ATP SYNTHASE

GENE STRUCTURE AND FUNCTION OF THERMOPHILIC ATP SYNTHASE

Yasuo Kagawa, Shigeo Ohta, Masafumi Yohda, Hajime Hirata,
Toshiro Hamamoto and Kakuko Matsuda

Department of Biochemistry, Jichi Medical School
Minamikawachi-machi, Tochigi-ken, Japan 329-04

ABSTRACT

An operon for thermophilic ATP synthase (TF_oF_1) was sequenced and mutated. The advantage of using TF_oF_1 for mechanistic studies is its reconstitutability without MgATP. All genes for TF_oF_1 were arranged in the order of promotors, structural genes coding for the I, TF_o subunits (a, c, b) and TF_1 subunits ($\delta, \alpha, \gamma, \beta$ and ϵ), and a terminator. The cause of the stability of these subunits was deduced from their sequence. The site-directed mutagenesis of the α and β subunits revealed that the 4 ionizable residues corresponding to Lys 21 and Asp 119 in the MgATP binding segments of adenylate kinase, are essential for the normal catalytic activity of this enzyme. The resulting βI164 and βN252 mutant subunits were both noncatalytic after reassembly into the $\alpha\beta\gamma$ subunit complex, even though both subunits bound significant amounts of ADP. The resulting αI175 reassembled weakly into an oligomer, while the αN261 was reconstituted into an $\alpha\beta\gamma$ subunit complex that showed no intersubunit cooperativity.

INTRODUCTION

ATP synthase (F_oF_1) catalyzes the formation of ATP from ADP and Pi by utilizing the electrochemical potential of protons (Mitchell, 1961). In fact, when F_oF_1 was incorporated into a planar lipid bilayer, the proton current could be directly measured (Hirata et al., 1986). The F_1 portion of the complex has catalytic activity and is bound to the F_o portion, which functions as a proton channel (Kagawa, 1984). F_1 consists of five different subunits $\alpha, \beta, \gamma, \delta$ and ϵ, of which α and β are homologous and bind AT(D)P even after their isolation. Although F_1 obtained from *E. coli* (EF_1)

Molecular Structure, Function, and Assembly of the ATP Synthases
Edited by Sangkot Marzuki
Plenum Press, New York

has been studied in detail (Futai and Kanazawa, 1983), it has been difficult to elucidate the MgAT(D)P binding site because both AT(D)P and Mg are strictly required for reassembly of the subunits into the $\alpha\beta\gamma$ complex (Dunn and Futai, 1980). F_1 obtained from the thermophilic bacterium PS3 (TF_1) is the only F_1 that can be reassembled without AT(D)P and Mg (Yoshida et al., 1977). Thus, it is possible to distinguish a mutant deficient in MgATP binding from one deficient in assembly (Table I). A study of the Cd, Mg-dependent diasteroisomer preference showed that its true substrate is a Δ, β, γ bidentate MgATP complex (Senter et al., 1983), which is the same as the substrate for adenylate kinase.   X-ray crystallography and NMR spectrometry of adenylate kinase revealed MgATP-binding segments containing Lys 21 and Asp 119 (Fry et al., 1986) which are homologous to the segments of both the α (Lys 175 and Asp 261) and β (Lys   164 and Asp 252) subunits of TF_1.   Thus, site-directed mutagenesis of these 4 residues in TF_1 was attempted.

MATERIALS AND METHODS

Over-expression of the Genes and Assay of the Products

TF_1 and its subunits were purified, reassembled and assayed by essentially the same methods as described previously (Kagawa and Yoshida, 1979).   The manipulation of thermophilic DNA has also been described in the previous reports (Kagawa et al., 1984, 1986 and 1987; Saishu et al., 1986).   The multifunctional plasmid pTZl8R/19R and host E. coli MN522 were obtained from Pharmacia, Sweden. pTZ was inserted with the genes (wild-type and mutated) for the α, β and γ subunits and manipulated as described elsewhere (Davis et al., 1986).   ADP binding to the subunits was determined by equilibrium dialysis and by difference UV absorption (Hisabori et al. 1986).

Site-Directed Mutagenesis

Four primers (21mers) for αI175, αN261, βI164 and βN252 were synthesized, in which the mismatch was introduced at the center of the oligonucleotide by changing the Lys or Asp codons into the Ile or Asn codons, respectively. The resulting 21mers were phosphorylated at their 5'-terminal and then incubated with single stranded templates (pTZl8R) to form the heteroduplex by polymerization and ligation. Competent E. coli MN522 cells were transformed with the heteroduplex, and 36 colonies produced on plates were isolated and infected with M13 KO7 helper phage. The single stranded DNA secreted into the medium was collected and sequenced to screen the colonies.

RESULTS AND DISCUSSION

Gene Structure of TF_oF_1

In the *Dra*I-*Sal*I fragment (13 kbp) of the thermophilic DNA, all genes for TF_oF_1 (8 kbp) were found in the order of promotors, structural genes coding for I protein, TF_o subunits (*a, c and b*), TF_1 subunits (α, β, γ, δ and ε), and terminator. All intercistronic non-coding regions of TF_oF_1, especially those of the *a-c* and *c-b* regions (Fig. 1), were longer than those of the *E. coli unc* operon. The most abundant codons used in TF_oF_1, which were rarely used in EF_oF_1 were TTG (Leu), GTC (Val), TCG (Ser), ACG (Thr) and Arg (CCG). Since the genes for Tα, Tβ and Tγ inserted in pTZ were expressed in large quantities in *E. coli* MN522 (Tα occupied 90% of total supernatant proteins), the Futai's hypothesis that subunit stoichiometry is determined by the codon usage, was denied.

Subunit Structure of TF_oF_1

TF_o was oligomycin insensitive because essential residues in both the *oli 2* and *oli 4* loci of the TF_o-*a* subunit (Fig. 2) were missing. TF_o-*a* was homologous to that of *Synechococcus* 6301 (Cozens and Walker, 1987). Although there was a long hydrophilic helix structure in TFo-*b*, the long helical structure reported in Eδ was not detected in Tδ which was homologous to mitochondrial F_o-OSCP (Fig. 3). The molecular weights and properties of TF_1 subunits are summarized in Table I. The external polarity, internal hydrophobicity, and propensities to form higher structures of the β and ε subunits were high in TF_oF_1 (Saishu et al., 1986; Kagawa et al., 1986). The primary structure of Tα, Tβ and Tγ were significantly more homologous to that of mitochondrial F_1 than that of EF_1. Rossmann fold reported by Futai on Eβ was not found in Tβ.

a-c TGACCATTGAGCATTCCCTTATCATTTCATTAAGGACTTGAAACATTACTTTAACATGGTGAAGGAGGATC
TATCAACATG

c-b TAAATCGATATGGAAAGTGATGGCGAAGTTCGTGTCAACGACCACTCGCCATTCCTTTATGTGCGGTTTTG
CGAGACCGCATCGGTGCGATACGTGTCGAAAGGAGTGAAACGCGGTGTTGTGGAAGGCAAACGTATGGGTG

b-δ TGATCGCAGCGTACATCAAAGACGTTCAAGAGGCAGGAGGAACGCGATG

δ-α TAACATTTGAAGACAGGGGTGAAAGGCATG

α-γ TAAGGCCGGCGGGGGCATATCTGCCCTCATGCCTGTTAGAACCCGTGCAGCAAAAAGGTGGTGAAACCTTTG

γ-β TAGGGCACTAGCAAGTTAGGAGGGAAAACGATG

β-ε TGACCCGGGATAGGGGGATTGGACAATG

Fig. 1. Intercistronic Non-coding Regions in the TF_oF_1 Operon.

```
oli 2 site          *0--*0#--00-0---0*00*****00***00000
PS3 TFo subunit a   PLK-IIEEFANTL---TLGLRLFGNIYAGEILLGL
Synechococcus  a    PFK-ILEDFTKPL---SLSFRLFGNILADELVVAV
Pea chloroplast a   PIN-IIEDFTKPL---SLSFRLFGNILADELVVAV
E.coli subunit a    PVNLILEGVSLLSKPVSLGLRLFGNMYAGELIFIL
Bovine Mt ATPase 6  PMLVIIETISLLIQPMALAVRLTANITAGHLLIHL
S.cerv.Mt ATPase 6  PLLVIMETLSYIARAISLGLRLGSNILAGHLLMVI


oli 4 site          0*0*00-*0--#-----**00*00*000*
PS3 TFo subunit a   AFSIFVG-T--------IQAFIFTMLTMVY
Synecoccus 6301 a   LFT-----S-------AIQALIFATLAASY
Pea chloroplast a   LFT-----S-------G-IQALIFATLAAAY
E.coli subunit a    IFHILI-IT--------LQAFIFMVLTIVY
Bovine Mt ATPase 6  TFTILILLTILEFAVAMIQAYVFTLLVSLY
S.cerv.Mt ATPase 6  PLAMILAIMILEFAIGIIQSYVWTILTASY
```

Fig. 2. Homology of TFo-*a* Subunit at Oligomycin Sensitive Loci. #: essential glutamate residues. *: mainly conserved residues. O: hydrophobic residues.

Site Directed Mutagenesis

ADP bound to TF_1 was synthesized into ATP-TF_1 without external energy in the presence of Pi and organic solvents (Yohda et al., 1986). Use of the TF_1 gene enabled us to distinguish the catalytic activity (lost in the $\beta I164$ and $\beta N252$) from the ability to reassemble without MgAT(D)P (lost only in $\alpha I175$), and to bind [^{14}C]ADP (partially conserved in $\alpha N261$, $\beta I164$ and $\beta N252$). A red-shift of absorption maximum of these subunits indicated the bound adenine portion surrounded by a hydrophobic

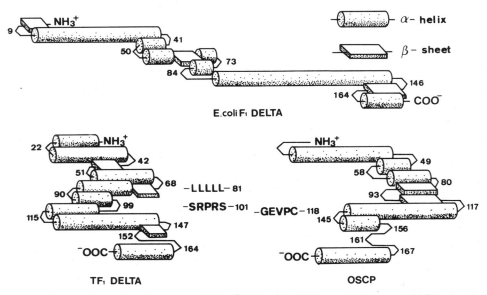

Fig. 3. Secondary Structure of the δ Subunit of TF_1 and EF_1, and OSCP (Oligomycin Sensitivity Conferring Protein of Beef Heart Mitochondria).

Table I. Properties of Thermophilic F_1 and *E. coli* F_1.

	TF$_1$	EF$_1$
Reassembly of $\alpha\beta\gamma$		
ATP requirement	none	strict
Mg requirement	none	strict
Active oligomers	$\alpha\beta\gamma,\alpha\beta\delta$	$\alpha\beta\gamma$
Renaturation of subunits	easy	impossible
Ligand binding		
removal of Mg–ATP	possible	denatured
ITP binding	only to β	?
stereochemistry	Δ,β,γ MgATP	unknown
Cd–ATP	active	inactive
Differentiation of mutants		
catalytic mutant vs. assembly mutant	possible	confusing
catalytic mutant vs. binding mutant	possible	confusing
catalytic mutant vs. ts–mutant	good	poor
Physical measurements		
NMR, IR, CD etc.	stable	difficult
Molecular weights (residues)		
α	54,590 (502)	55,259 (513)
β	51,938 (473)	50,117 (459)
γ	31,706 (282)	31,414 (287)
δ	19,683 (179)	19,303 (177)
ϵ	14,333 (132)	14,914 (138)

environment (Hisabori et al., 1986). In fact, the ATP binding hydrophobic residues (Ile28 and Val29) of adenylate kinase (Fry et al., 1986) were also conserved in TF$_1$. One role of the lysine containing segment in the β subunit is assumed to be relocation of catalytic groups towards the reaction center of the bound substrate. ADP binding activity of the mutant subunits ruled out that the loss of catalytic activity was caused by an nonspecific denaturation of the mutant subunits. The αN261-β-γ complex showed ATPase activity (Km = 20 μM and Kcat = 6 sec^{-1}) which was close to that of the wild-type at low ATP concentrations, but when the ATP concentration was increased, there was no enhancement in both Km and Vmax as in the case of the wild-type. The loss of cooperativity in the catalytic activity of the αN261-β-γ complex and the decreased ability to reassemble in αI175 suggested that the mutated residues of the α subunit have a role in the inter-subunit interaction.

ACKNOWLEDGMENT

This work was supported in part by Grant-in-Aid for Scientific Research on Priority Areas of "Bioenergetics" to Y.K. from the Ministry of Education, Science and Culture, Japan.

REFERENCES

Cozens, A.L. and Walker, J.E. (1987) J. Mol. Biol. 194, 359-383.

Davis, L.G., Dibner, M.D., and Battey, J.F. (1986) Basic Methods in Molecular Biology, Elsevier, Amsterdam.

Dunn, S.D. and Futai, M. (1980) J. Biol. Chem. 255, 113-118.

Fry, D.C., Kuby, S.A. and Mildvan, A.S. (1986) Proc. Natl. Acad. Sci. USA 83, 907-911.

Futai, M., and Kanazawa, H. (1983) Microbiol. Rev. 47, 258-312.

Gay, N.J. and Walker, J.E. (1981) Nucleic Acids Res. 9, 3919-3926.

Hirata, H., Ohno, K., Sone, N., Kagawa, Y. and Hamamoto, T. (1986) J. Biol. Chem. 261, 9839-9843.

Hisabori, T., Yoshida, M., and Sakurai, H. (1986) J. Biochem. 100, 663-670.

Kagawa, Y. (1984) in Bioenergetics (Ernster, L. Ed.) pp. 149-186, Elsevier, Amsterdam.

Kagawa, Y., Hirata, H., Ohta, S., Ishizuka, M. and Karube, Y. (1987) in Ion Transport through Membranes (Yagi, K. and Pullman, B. Eds.) pp. 147-162, Academic Press, New York.

Kagawa, Y., Ishizuka, M., Saishu, T., and Nakao, J. (1986) J. Biochem. 100, 923-934.

Kagawa, Y., Nojima, H., Nukiwa, N., Ishizuka, M., Nakajima, T., Yasuhara, T., Tanaka, T. and Oshima, T. (1984) J. Biol. Chem. 259, 2956-2960.

Kagawa, Y. and Yoshida, M. (1979) Methods Enzymol. 55, 781-787.

Mitchell, P. (1961) Nature 191, 144-148.

Saishu, T., Nojima, H. and Kagawa, Y. (1986) Biochim. Biophys. Acta 867, 97-106.

Senter, P., Eckstein, F., and Kagawa, Y. (1983) Biochemistry 22, 5514-5518.

Yohda, M., Kagawa, Y., and Yoshida, M. (1986) Biochim. Biophys. Acta 850, 429-435.

Yoshida, M., Sone, N., Hirata, H., and Kagawa, Y. (1977) J. Biol. Chem. 252, 3480-3485.

GENETIC STUDIES OF F_1-ATPase OF *ESCHERICHIA COLI*

Masamitsu Futai, Takato Noumi and Masatomo Maeda

Department of Organic Chemistry and Biochemistry, The Institute of Scientific and Industrial Research, Osaka University, Osaka 567, Japan

I. INTRODUCTION

Escherichia coli ATP synthase (H^+-ATPase) is the first energy transducing enzyme whose primary structure has been determined from the DNA sequence of the cloned genes (*unc* operon) (Kanazawa and Futai, 1982; Futai and Kanazawa, 1983; Walker et al., 1984; Senior, 1985). This enzyme (F_oF_1) is similar to those found in mitochondria or chloroplasts and the catalytic entity F_1 (F_1-ATPase) consists of five subunits α, β, γ, δ and ϵ. Studies of *E. coli* F_1 are advantageous because variant (mutant) enzymes with defined amino acid substitutions can be easily obtained. Studies on such enzymes may help to understand mechanism and assembly of the normal enzyme. Furthermore most of the results can be extended to the eukaryotic enzymes. This short article summarizes our recent results on the genetic studies of F_1-ATPase.

II. ASSEMBLY MUTATIONS AND SUBUNIT-SUBUNIT INTERACTIONS

In Table I, we have listed *unc⁻* mutants which mutation sites were determined in our laboratory. Mutation often causes defective assembly of the entire complex. Replacements in Glu-41, Glu-185, Ser-292 and Gly-223 of the β subunit resulted in assembly mutations (Noumi et al., 1986a). It is noteworthy that these residues are conserved in β subunits from other organisms. All assembly mutants had reduced amounts of α and β subunits, suggesting that these residues or the region in their vicinities are important in interactions of the two subunits. More interestingly, some mutant membranes lost the ϵ subunit (Glu-41 → Lys), δ and ϵ subunits (Glu-185 → Lys), or γ and ϵ subunits (Ser-292 → Phe), although their membranes had α and β

Molecular Structure, Function, and Assembly of the ATP Synthases
Edited by Sangkot Marzuki
Plenum Press, New York

Table I. Mutations in F_1 Subunits.

Defect	Strain	Subunit	Codon Replaced	Residue[a] Replaced	Reference
I. Assembly	KF39	β	GAA(42) →AAA	Glu-41 →Lys	Noumi et al., 1986a
	KF16 KF42	β	GAG(186)→AAG	Glu-185→Lys[b] Glu-185→Gln[b]	Noumi et al.[1]
	KF48	β	CCT(224)→GAT	Gly-223→Asp	Noumi et al., 1986a
	KF26 KF27 KF30 KF32 KF37	β	TCC(293)→TTC	Ser-292→Phe	”
	KF40	β	CAG(362)→TAG	Gln-361→End	”
	KF20	β	CAG(362)→TAG	Gln-397→End	,,
	KF122	γ	ATG(1) →ATA		Miki et al.[c]
	KF10	γ	CAG(15) →TAG	Gln-14 →End	Miki et al., 1986
	KF1 KF68 KF88	γ	CAG(158)→TAG	Gln-157→End	,,
	KF21	γ	CAG(227)→TAG	Gln-226→End	,,
	KF84	γ	CAG(262)→TAG	Gln-261→End	
	KF110	γ	CAG(230)→TAG	Gln-229→End	Miki et al.[c]
	KF12 KF13	γ	CAG(270)→TAG	Gln-269→End	Miki et al., 1986
	NR70	γ	Δ 22~28	Δ 21~27[d]	Kanazawa et al., 85
	KF148	ε	CAG(73) →TAG	Gln-72 →End	Kuki et al.[e]
	KF53	ε	GGAG→AAAG[f]		,,
II. Catalysis	uncA401	α	TCC(373)→TTC	Ser-373→Phe	Noumi et al., 1984a
	KF101	α	GCA(285)→GTA	Ala-285→Val	Soga et al.[g]
	KF114	α	CGT(376)→TGT	Arg-376→Cys	,,
	KF11 KF168	β	TCT(175)→TTT	Ser-174→Phe	Noumi et al., 1984b
	KF43	β	CGT(247)→CAT	Arg-246→His	Noumi et al., 1986b
	KF104	β	CGT(247)→TGT	Arg-246→Cys	Kuki et al.[h]
	KF87	β	GCG(152)→GTG	Ala-151→Val	Hsu et al., 1987

Mutations of F_1 identified in our laboratory are summarized. Manuscripts in preparation will be submitted in a few months. [1]Noumi, T., Azuma, M., Shimomura, S., Maeda, M. and Futai, M., in press. [a]As indicated previously (Kanazawa et al., 1981; Dunn, 1982), the isolated β and γ subunits do not have Met residues at the amino terminus, and thus residues are numbered from the Ala residue (2nd codon for both subunits). [b]Mutant β subunit isolated by applying site-directed mutagenesis: T. Noumi, M. Azuma, S. Shimomura, M. Maeda and M. Futai, in preparation. [c]J. Miki, M. Maeda and M. Futai, in preparation. [d]A deletion of 7 amino acid residues. [e]M. Kuki, T. Noumi, M. Maeda and M. Futai, in preparation. [f]Mutation in Shine-Dargarno sequence. [g]S. Soga, T. Noumi, M. Maeda, M. Futai, in preparation. [h]M. Kuki, T. Noumi, M. Maeda, M. Futai, in preparation.

subunits. These results cast doubts on a conceptual model of F_1 in which $\alpha\beta\gamma$ complex is connected to F_o subunits through δ and ϵ subunits. Thus the model of F_oF_1 should include direct interaction(s) between catalytic (α and β) subunits and F_o subunits as indicated by the results of cross-linking experiments (Aris and Simoni, 1983). The importance of Glu-185 or the region in its vicinity in assembly was confirmed by applying site directed mutagenesis to this residue (Noumi, T., Azuma, M., Shimomura, S., Maeda, M. and Futai, M., in press). The purified β subunit with replacement of Glu-185 by Gln or Lys could not form a complex with α and γ subunits *in vitro*. The mutant β subunits had altered binding activities for aurovertin and stabilities of the secondary structures.

Mutations in γ subunit also caused assembly defect of the entire complex (Kanazawa et al., 1985; Miki et al., 1986). Only negligible amounts of α and β subunits were found in membranes of the mutant lacking 16 residues (Gln-260 → end) in carboxyl terminal region (Miki et al., 1986) or the strain carrying a deletion (Δ 21-27) in amino terminal region (Kanazawa et al., 1985). The both regions are highly conserved in the γ subunit so far sequenced, and may be important for assembly of the entire complex. The importance of the γ subunit in assembly is consistent with the reconstitution experiments: the ATPase complex could be reconstituted by a combination of the α, β and γ subunits, but not without the γ subunit (Futai, 1977; Dunn and Futai, 1980).

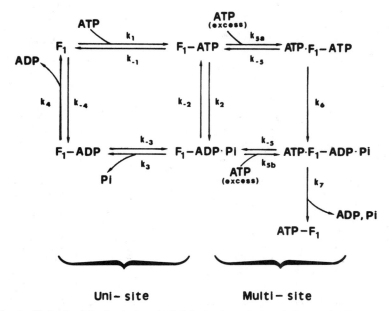

Fig. 1. Uni-site (single site) and Multi-site (steady state) Catalysis of F_1. The scheme of ATP hydrolysis (and synthesis) by F_1 is cited and modified from Penefsky and coworkers (Grubmeyer et al., 1982; Cross et al., 1982).

Table II. Catalytic Properties of F_1 from Wild-type and Mutants of *E. coli*

Parameter	F_1	F_1+NaN$_3$ [a]	AP$_3$-PL.F$_1$ [b]	KF43(Arg-246→His)F$_1$ [c]	KF87(Ala-151→Val)F$_1$ [d]
$k_1(M^{-1}S^{-1})$	3×10^4	2×10^4	$\fallingdotseq 8 \times 10^2$	2×10^4	$\geq 33 \times 10^4$
$k_{-1}(S^{-1})$	$\leq 6 \times 10^{-5}$	$\leq 6 \times 10^{-5}$	2×10^{-4}	ND[e]	ND[e]
$F_1.ATP/F_1.ADP.Pi$	2:1	3:2			
$k_3(S^{-1})$	1.8×10^{-3}	1.4×10^{-3}	~ 0[f]	$\geq 3 \times 10^{-2}$	50~100 fold higher than wild-type F_1
$k_7(S^{-1})$ (Multi-site)	78	6.2	~ 0[f]	1.9	4.6
k_7/k_3 (Promotion of catalysis)	4.3×10^4	4.4×10^3	-	≤ 63	≤ 46

[a]Kinetic parameters of wild-type F_1 with or without NaN$_3$; cited from Noumi et al., 1987a.
[b]Kinetic parameters of wild-type F_1 modified with AP$_3$.PL; cited from Noumi et al., 1987b.
[c]Cited from Noumi et al., 1986b.
[d]Cited from Hsu et al., 1987.
[e]The k_{-1} could not be determined because the amounts of ATP and Pi bound to F_1 were too low.
[f]Modified F_1 is suggested to have no uni- and multi-site activities.

No real missense mutations were found in γ subunit, suggesting that amino acid replacements in this subunit, if any, did not result in a serious defects in function and assembly of the complex and that the numbers of essential amino acid residues are limited. Defects in assembly seem to occur with mutations in other subunits, although no detailed studies were made; mutations in the α subunit caused defective solubilization of F_1 upon washing with dilute buffer (unpublished observation).

III. MUTATIONS IN CATALYTIC MECHANISM

1. Uni- and Multi-site Catalyses

The cooperative mechanism of ATP hydrolysis by F_1 from mitochondria (Cross et al., 1982; Grubmeyer et al., 1982) and later from *E. coli* (Duncan and Senior, 1985; Noumi et al., 1986b) has been studied extensively; the ATP at the first catalytic site is hydrolyzed slowly (uni-site hydrolysis), but on binding of ATP at the second and third sites, the ATP at the first site is hydrolyzed (with release of Pi and ADP) at maximal velocity (multi-site or steady state hydrolysis) (Fig. 1). The ratio of the multi- and uni-site rate is 10^6 for mitochondrial F_1 and 10^4 - 10^5 for *E. coli* F_1. Sodium azide inhibited multi-site hydrolysis by *E. coli* F_1, whereas it did not affect uni-site ATPase activity (Table II) (Noumi et al., 1987a). Thus azide inhibited multi-site hydrolysis by lowering catalytic cooperativity. Consistent with this interpretation, azide changed the ligand-induced fluorescence response of aurovertin bound to F_1.

F_1 modified with about one mole of adenosine triphosphopyridoxal (AP$_3$-PL) had essentially no uni- and multi-site hydrolysis of ATP (Table II) (Noumi et al., 1987b). The rate of ATP binding (k_1) decreased to 10^{-2} of that of the unmodified F_1 and the rate of release of ATP (k_{-1}) was about two times faster. The equilibrium $F_1.ATP \rightleftharpoons F_1.ADP.Pi$ was shifted toward $F_1.ATP$. These results suggest that the ATP analoque bound to an active site, and catalysis by the two remaining sites was completely abolished. These results support the model in which the three catalytic sites are proposed to undergo sequential conformational changes.

2. Mutations in α and β Subunits Affect Multi-site Catalysis

Mutations in both α and β subunits resulted in low multi-site activity. The purified mutant F_1 from *uncA401* with replacement of α.Ser-373 → Phe (Noumi et al., 1984a) showed low multi-site hydrolysis, but a normal uni-site hydrolysis (Wise et al., 1984; Kanazawa et al., 1984). Pi-water exchange experiments suggested that the equilibrium $F_1.ATP \rightleftharpoons F_1.ADP.Pi$ is shifted toward $F_1.ATP$ (Wood et al., 1987).

The F_1 from strain KF43 (β.Arg-246 → His) has low multi-site activity (k_7) (1-3% of the wild-type F_1) (Noumi et al., 1986b). The k_3 (the rate of Pi release in uni-site hydrolysis) was about 15 fold more than that of wild-type F_1, although wild-type and mutant F_1 had similar k_1 values (rate of ATP binding in uni-site) (Table II). It is noteworthy that a single amino acid replacement altered both k_3 and k_7. We also isolated a mutant (β.Arg-246 → Cys) which F_1 had similar kinetic properties to the KF43 enzyme. F_1-ATPase from strain *uncD484* (β.Met-209 → Ile) (Duncan et al., 1986) had also altered kinetics in uni-site and multi-site hydrolysis (Duncan and Senior, 1985). Thus the both residues or regions in their vicinities may be related to Pi release steps in uni-site and positive cooperativity for multi-site hydrolysis. As discussed above, single amino acid replacements either in α or β subunit resulted in loss of cooperativity and multi-site catalysis. For the positive cooperativity of catalysis, conformational transmission between different catalytic sites is essential, and the pathway of the transmission may be formed from amino acid residues in α and β subunits.

It was interesting to know whether F_1-ATPase was still active in multi-site catalysis when one of its β subunits was defective in conformational transmission. To answer this question, we constructed hybrid $\alpha\beta\gamma$ complex (ATPase) having different ratios of β subunits of the wild-type and a mutant KF43. Analysis of the experimental results showed that only the complex reconstituted with wild-type β subunit was active, whereas complexes containing one or two mutant β subunits were not active (Noumi et al., 1986b). Thus all three β subunits must be competent in conformational transmission for the cooperativity which is required for multi-site ATPase activity.

The F_1 from strain KF11 (β.Ser-174 → Phe) (Noumi et al., 1984b) was different from those of above mutants. It had substantial remaining multi-site ATPase activity with altered divalent cation dependency: the ratio of Ca^{2+}- to Mg^{2+}- dependent multi-site activities in the mutant and wild-type F_1 were 3.5 and 0.8 (Kanazawa et al., 1980), respectively, suggesting that Ser-174 may be closely related to the Mg^{2+} binding site(s). The presence of a Mg^{2+} binding site in β subunit was indicated by studies using isolated subunit (Futai et al., 1987).

IV. AN AMINO ACID REPLACEMENT WITHIN A CONSERVED SEQUENCE

The sequence of the β subunit is highly conserved in different organisms and the sequence G-X-X-X-X-G-K-T/S-X-X-X-X-X-X-I/V (X can be any amino acid residue) is found in the β subunit of F_1, adenylate kinase, and other nucleotide binding proteins (Table III) (Walker et al., 1982; Futai and Kanazawa, 1983: Fry et al., 1985; Duncan et al., 1986). Residues 15-23 (G-G-P-G-S-G-K-G-T) of adenylate kinase form a flexible

Table III. Conserved Sequence of Nucleotide Binding Proteins.

Nucleotide Binding Protein	Sequence		References
Conserved sequence	G X X X X G K T/S		Kanazawa et al., 1982
E. coli F₁-β (wild)	G G A G V G K T	(149-156)	Kanazawa et al., 1982
(KF87)	G G V G V G K T	(149-156)	Hsu et al., 1987
E. coli F₁-α	G D R G T G K T	(169-178)	Kanazawa et al., 1981
Normal p21 ras	G A G G V G K S	(10-17)	Tabin et al., 1982 Reddy et al., 1982
Activated p21 ras	G A V G V G K S	(10-17)	Tabin et al., 1982 Reddy et al., 1982
Adenylate kinase (porcine)	G G P G S G K G T	(15-23)	Heil et al., 1974
EF-Tu (E. coli)	G H V D H G K T	(18-25)	Jones et al., 1980

The conserved sequence G-X-X-X-X-G-K-T/S-X-X-X-X-X-X-I/V has been found in more than 40 nucleotide binding proteins (Kanazawa et al., 1982; Walker et al., 1982; Futai and Kanazawa, 1983; Fry et al., 1985; Moller and Amons, 1985). Examples are shown in this Table together with the mutation of the β subunit found in this study.

loop structure between α helix and β sheet, and Lys-21 (K) may be close to the α or γ-phosphate of ATP (Fry et al., 1985; Tagaya et al., 1987). ATPase activity of the mitochondrial F₁ was lost by modification of the corresponding Lys with 7-chloro-4-nitrobenzofrazan (Andrews et al., 1984). Thus residues 149-156 (G-G-A-G-V-G-K-T) of the E. coli β subunit may have a similar role.

We have identified a mutation (strain KF87 Ala-151 → Val) within the conserved sequence (Hsu et al., 1987). The mutant enzyme had about 6% of the wild-type multi-site activity (Table II). Interestingly, the k_3 (rate of release of Pi in uni-site) value of the mutant enzyme was at least 50-100 fold higher than that of the wild-type enzyme (Table II). The mutant enzyme was defective in the transmission of conformational changes as studied by aurovertin fluorescence. Thus an amino acid replacement in the conserved sequence changed kinetics of the enzyme activity entirely, suggesting the importance of the sequence.

It is interesting to note that normal and oncogenic (activated) p21 ras protein have Gly and Val residues (Table III) (Tabin et al., 1982; Reddy et al., 1982),

15

respectively, at position corresponding to Ala-151 of the β subunit. Like the mutant F_1, the oncogenic protein had lower GTPase activity (about 10% of normal) (Sweet et al., 1984), but normal GTP binding activity. The side chain volume of Gly, Ala and Val residues are 66.1, 91.5 and 141.7 $\overset{\circ}{A}^3$, respectively (Chothia, 1975), and the replacement of Gly (p21 *ras*) or Ala (β subunit) by Val may change the structure of ATP binding site or the flexibility of the "loop" of the conserved region and orientation of Lys-16 (p21 *ras*) or Lys-157 (β).

V. CONCLUSION

In this article we have discussed our recent studies on F_1-ATPase of *E. coli*. Mutant enzymes with defined amino acid replacements have been obtained and their assembly and kinetics were studied. Structural and catalytic aspects of the enzyme deduced from these studies are discussed. Further studies on enzymes obtained from mutagenized cells or by site directed mutagenesis will lead to the detailed understanding of the structure and mechanism of the enzyme in molecular level.

ACKNOWLEDGMENTS

The studies in our laboratory discussed in this article were carried out in collaboration with Dr. H. Kanazawa, J. Miki, K. Takeda-Miki, and coworkers whose names appear in the references and was supported in part by a Grant-in-Aid from the Ministry of Education, Science and Culture of Japan, and Special Coordination Fund for Promotion of Science and Technology of the Science and Technology Agency of the Japanese Government. Support from the Mitsubishi Foundation is also gratefully acknowledged.

REFERENCES

Andrews, W.W., Hill, F.C. and Allison, W.S. (1984) J. Biol. Chem. 259, 14378-14382.

Aris, J.P. and Simoni, R.D. (1983) J. Biol. Chem. 258, 14599-14609.

Chothia, C. (1975) Nature 254, 304-308.

Cross, R. L., Grubmeyer, C. and Penefsky, H.S. (1982) J. Biol. Chem. 257, 12101-12105.

Duncan, T.M. and Senior, A.E. (1985) J. Biol. Chem. 260, 4901-4907.

Duncan, T.M., Parsonage, D. and Senior, A.E. (1986) FEBS Lett. 208, 1-6.

Dunn, S.D. and Futai, M. (1980) J. Biol. Chem. 255, 113-118.

Dunn, S.D. (1982) J. Biol. Chem. 257, 7354-7359.

Fry, D.C., Kuby, S.A. and Mildvan, A.S. (1985) Biochemistry 24, 4680-4694.

Futai, M. (1977) Biochem. Biophys. Res. Commun. 79, 1231-1237.

Futai, M. and Kanazawa, H. (1983) Microbiol. Rev. 47, 285-312.

Futai, M., Shimomura, S. and Maeda, M. (1987) Arch. Biochem. Biophys. 254, 313-318.

Grubmeyer, C., Cross, R.L. and Penefsky, H.S. (1982) J. Biol. Chem. 257, 12092-12100.

Heil, A., Muller, G., Noda, L., Pinder, T., Schirmer, H., Schirmer, I. and von Zaben, I. (1974) Eur. J. Biochem. 43, 131-144.

Hsu, S-Y., Noumi, T., Takeyama, M., Maeda, M., Ishibashi, S. and Futai, M. (1987) FEBS Lett., in press.

Jones, M.D., Petersen, T.E., Nielsen, K.M., Magnusson, S., Sottrup-J, L., Gausing, K. and Clark, B.F.C. (1980) Eur. J. Biochem. 108, 507-526.

Kanazawa, H., Horiuchi, Y., Takagi, M., Ishino, Y. and Futai, M. (1980) J. Biochem. 88, 695-703.

Kanazawa, H., Kayano, T., Mabuchi, K. and Futai, M. (1981) Biochem. Biophys. Res. Commun. 103, 604-612.

Kanazawa, H., Kayano, T., Kiyasu, T. and Futai, M. (1982) Biochem. Biophys. Res. Commun. 105, 1257-1264.

Kanazawa, H., Noumi, T., Matsuoka, I., Hirata, T. and Futai, M. (1984) Arch. Biochem. Biophys. 228, 258-269.

Kanazawa, H., Hama, H., Rosen, B.P. and Futai, M. (1985) Arch. Biochem. Biophys. 241, 364-370.

Miki, J., Takeyama, M., Noumi, T., Kanazawa, H., Maeda, M. and Futai, M. (1986) Arch. Biochem. Biophys. 251, 458-464.

Moller, W. and Amons, R. (1985) FEBS Lett. 186, 1-7.

Noumi, T., Futai, M. and Kanazawa, H. (1984a) J. Biol. Chem. 259, 10076-10079.

Noumi, T., Mosher, M.E., Natori, S., Futai, M. and Kanazawa, H. (1984b) J. Biol. Chem. 259, 10071-10075.

Noumi, T., Oka, N., Kanazawa, H. and Futai, M. (1986a) J. Biol. Chem. 261, 7070-7075.

Noumi, T., Taniai, M., Kanazawa, H. and Futai, M. (1986b) J. Biol. Chem. 261, 9196-9201.

Noumi, T., Maeda, M. and Futai, M. (1987a) FEBS Lett. 213, 381-384.

Noumi, T., Tagaya, M., Maeda, M., Takeda-Miki, K., Fukui, T. and Futai, M. (1987b) J. Biol. Chem., in press.

Reddy, E.P., Reynolds, R.K., Santos, E. and Barbacid, M. (1982) Nature 300, 149-152.

Senior, A.E. (1985) Cur. Top. Membr. Transport. 23, 135-151.

Sweet, R.W., Yokoyama, S., Kamata, T., Feramisco, J.R., Rosenberg, M. and Gross, M. (1984) Nature 311, 273-275.

Tabin, C.J., Bradley, S.M., Bargmann, C., Weinberg, R. A., Papageorge, A.G., Scolnick, E.M., Dhar, R., Lowy, D.R. and Chang, E.H. (1982) Nature 300, 143-149.

Tagaya, M., Yagami, T. and Fukui, T. (1987) J. Biol. Chem., in press.

Walker, J.E., Saraste, M., Runswick, M.J. and Gay, N.J. (1982) EMBO J. 1, 945-951.

Walker, J.E., Saraste, M. and Gay, N.J. (1984) Biochim. Biophys. Acta 768, 164-200.

Wise, J.G., Latchney, L.R., Ferguson, A.M. and Senior, A.E. (1984) Biochemistry 23, 1426-1432.

Wood, J.M., Wise, J.G., Senior, A.E., Futai, M. and Boyer, P.D. (1987) J. Biol. Chem. 262, 2180-2186.

THE CHLOROPLAST GENES ENCODING CF_O SUBUNITS OF ATP SYNTHASE

Graham S. Hudson, John G. Mason, Tim A. Holton, Paul Whitfeld, Warwick Bottomley and Graeme B. Cox*

CSIRO, Division of Plant Industry, and *Biochemistry Department John Curtin School of Medical Research, Australian National University, Canberra, ACT, Australia

The H^+-ATP synthase of the chloroplast thylakoid membranes is essential for electron transport and photophosphorylation during photosynthesis. It has an F_1F_0 structure analogous to the proton-translocating ATP synthases of bacteria and mitochondria, that is, a proton pore sector (CF_O) within the thylakoid membrane and an ATPase moiety (CF_1) protruding into the stroma. In this paper, we examine the composition and predicted structure of the CF_O as deduced from three genes for CF_O subunits and from data for the equivalent proteins from *E. coli* and mitochondria.

SUBUNIT COMPOSITION OF THE CHLOROPLAST ATP SYNTHASE

The complex is thought to be composed of eight or nine subunit types (Table I). The genes for three of the five CF_1 subunits (alpha, beta, epsilon) are on the chloroplast genome at two loci, and encode polypeptides homologous to subunits of the same name in the *E. coli* F_1. The gamma and delta subunits are encoded in the nucleus as precursor polypeptides of higher molecular weight (Watanabe and Price, 1982). Usually, three CF_O subunits designated I, II and III can be detected by polyacrylamide gel electrophoresis, although an additional band of 19 kDa (called X) was resolved in gels containing urea (Pick and Racker, 1979; Westhoff et al., 1985). Subunit III is the DCCD-binding proteolipid equivalent to the subunit c of *E. coli* and subunit 9 of mitochondria. Suss and Schmidt (1982) have estimated the composition of the CF_1CF_0 as $alpha_3$ $beta_3$ gamma delta epsilon I II III_5, while the *E. coli* F_1F_0 has the probable composition of $alpha_3$ $beta_3$ gamma delta epsilon a b_2 c_{6-10} (Sebald and Hoppe, 1981; Foster and Fillingame, 1982). The genes for subunits I and III were mapped to the chloroplast genome near the gene for the alpha subunit, *atp*A, while

Molecular Structure, Function, and Assembly of the ATP Synthases
Edited by Sangkot Marzuki
Plenum Press, New York

Table I.   Composition of the Chloroplast H^+-ATP Synthase from Spinach.

Subunit	Apparent M.W.[a] (kDa)	No. Per Complex[b]	Gene Name	Gene Location	Predicted M.W.[c] (kda)	E.coli Subunit Homologue
CF_1:						
alpha	59	3	atpA	chloroplast	56	alpha
beta	52	3	atpB	chloroplast	54	beta
gamma	37	1	atpC	nucleus		
delta	20	1	atpD	nucleus		
epsilon	16	1	atpE	chloroplast	15	epsilon
CF_0:						
I	18	1	atpF	chloroplast	$21(19^d)$	b
II	16	1	atpG	nucleus		
III	8	5	atpH	chloroplast	8	c
IV	19		atpI	chloroplast	27	a

[a]From SDS gel electrophoresis data (Westhoff et al., 1985).
[b]For *Vicia faba* and *Avena sativa* (Suss and Schmidt, 1982).
[c]From nucleotide sequence data (Zurawski et al., 1982; Hudson et al., 1987).
[d]Processed molecular weight.

subunit II was found by *in vitro* translation of poly(A)$^+$ RNA to be nuclear encoded as a 24 kDa precursor protein (Westhoff et al., 1985).   Subsequent DNA sequencing of the pea and spinach alpha operons revealed the presence of three, not two, CF_0 genes associated with atpA (Cozens et al., 1986; Hennig and Herrmann, 1986; Hudson et al., 1987) so that the composition of the CF_0 needs to be reconsidered.

PREDICTED STRUCTURES OF THE CF_0 SUBUNITS

The alpha operon contains four genes, atpIHFA (Fig. 1), with the most distal gene encoding the alpha subunit which is 54% identical in amino acid sequence to the alpha subunit of *E. coli*.   The next gene upstream is atpF which is split by a single intron.   The exact splice sites have been determined by isolation of a cDNA for the spinach gene (Hudson et al., 1987).   The amino terminal sequence of spinach subunit I has been shown to correspond to the atpF reading frame from residue 18 onwards, providing evidence that the first 17 residues are removed post-translationally (Bird et al., 1985). In pea, the leader peptide is only 6 residues in length so that the sequence requirements for the leader peptide are apparently not strict. Subunit I features a predicted two domain structure (Fig. 2A) with an N-terminal membrane anchorage segment and the remainder of the molecule as a hydrophilic stalk of two opposing alpha-helices protruding into the stroma. Although the amino acid sequence of subunit I is only poorly homologous (19%) to subunit b of $E.$ $coli$ F_o, it is apparent that the secondary structure of the two polypeptides is similar. The hydrophilic domains of the two b subunits of the $E.$ $coli$ F_o have been proposed as the structure to which the F_1 is attached (Perlin et al., 1983; Walker et al., 1984).

The atpH gene encodes the proteolipid subunit III of CF_o and the predicted amino acid sequence from the spinach gene exactly matches that determined for the purified protein (Sebald and Wachter, 1980). The predicted secondary structure for subunit III or its $E.$ $coli$ homologue, subunit c, is two membrane-spanning alpha-helices connected by a reverse turn (Fig. 2B). The chloroplast protein is 29% identical to subunit c, and features buried proline and glutamic acid residues in the first and second alpha-helices respectively while the $E.$ $coli$ protein has buried proline and aspartic acid residues both in its second alpha-helix. The aspartic acid of subunit c is the site of interaction with DCCD and thus is crucial to proton pore function (Sebald and Hoppe, 1981). Fimmel et al. (1983) have demonstrated that mutation of the proline leads to loss of ATPase function which can be rescued by second site revertants containing a proline in the first alpha-helix. Bending of the alpha-helix by the proline residue may be important for the spatial packing of the proteolipid subunits into F_o.

The fourth and most proximal gene, atpI, encodes a 27 kDa hydrophobic polypeptide whose amino acid sequence shows limited homology to subunit a of $E.$ $coli$ and to subunit 6 of mitochondrial ATPase, mostly towards their carboxy termini. Based on models for the $E.$ $coli$ and mitochondrial subunits (Cox et al., 1986) the predicted secondary structure for the atpI protein shows five transmembrane alpha-helices with the fourth having an amphipathic character (Fig. 2C). The proton pore is proposed to be formed by the interaction of residues on one face of the fourth helix of

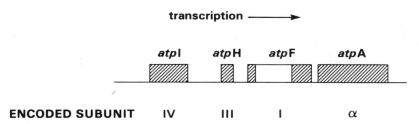

Fig. 1. Organization of the atpIHFA Operon in the Spinach Chloroplast Genome.

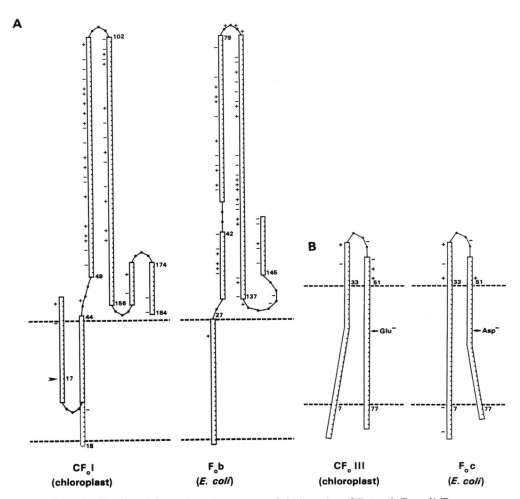

Fig. 2. Predicted Secondary Structures of Chloroplast CF_o and *E. coli* F_o Subunits. (A) CF_oI and F_ob. (B) CF_oIII and F_oc. (C) CF_oIV and F_oa. The dotted lines represent the outer surfaces of the chloroplast thylakoid or *E. coli* cytoplasmic membrane.

C

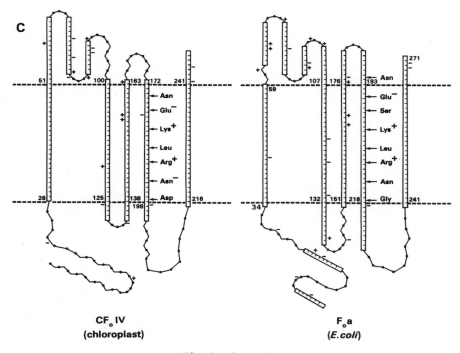

CF$_o$ IV
(chloroplast)

F$_o$a
(E.coli)

Fig. 2.   (cont.)

SUBUNIT	E. coli.		SPINACH CHLOROPLAST		BOVINE MITOCHONDRION	
	a	c	a (IV)	c (III)	a (6)	c (9)
	Glu$^-_{196}$		Asn$_{175}$		Glu$^-_{145}$	
	Ser		Glu$^-_{178}$		Ser	
	Lys$^+$		Lys$^+_{182}$		Gln	
	Leu	$^-$Asp$_{61}$	Leu$_{186}$	$^-$Glu$_{61}$	Leu	$^-$Glu$_{58}$
	Arg$^+$		Arg$^+_{189}$		Arg$^+$	
	Asn		Asn$_{193}$		Asn	
	Gly		Asp$^-_{197}$		Gly	

Fig. 3. Comparison of the Amino Acid Residues on the Amphipathic Face of the Fourth Alpha-helix (see Fig. 2c) of Subunits a, IV and 6 of the *E. coli*, Spinach Chloroplast and Bovine Mitochondrial ATP synthases. These residues interact with the glutamic or aspartic acid of subunit c, III or 9 to form the proton pore (Cox et al., 1986).

23

the a-type subunit with the buried aspartic or glutamic acid of the c-type subunit, and it is these residues which are quite conserved between the bacterial, chloroplast and mitochondrial subunits (Fig. 3). Some support for this model comes from the isolation of mutations in the fourth and fifth helices of subunit a which affect proton translocation in the *E. coli* ATPase (Cain and Simoni, 1986). Also conserved in this fourth helix between chloroplast, *E. coli* and mammalian (but not yeast) mitochondrial subunits is a proline residue which would cause a bend in the helix.

FOUR CF$_0$ SUBUNITS?

It seems clear that *atp*I encodes a subunit of the CF$_0$ and is actively transcribed *in vivo*. Does this gene product correspond to the 16 kDa subunit II or the 19 kDa band X observed by Westhoff et al. (1985)? The evidence for a nuclear-encoded subunit II is strong, and Cozens et al. (1986) have demonstrated in a coupled transcription-translation system that *atp*I gives rise to a 20 kDa product, close to the apparent size of X. Formal biochemical proof will be required to demonstrate the presence of *atp*I protein within the CF$_0$ as a new subunit, IV, but indirect evidence from the sequencing of ATPase genes in the cyanobacterium *Synechococcus* 6301 has provided a basis for the concept of four CF$_0$ subunits (Cozens and Walker, 1987). The organization of the cyanobacterial *atp* genes as two operons, with the genes for the beta and epsilon subunits separated from the others, and the close homology of the predicted cyanobacterial subunits to the chloroplast subunits supports the endosymbiont hypothesis that the ancestor of the chloroplast was a photosynthetic bacterium engulfed by a eukaryotic cell. Most importantly, two genes for b-type subunits (b,b')are found within the first operon, so that the cyanobacterial F$_0$ may contain one copy of each subunit whereas *E. coli* F$_0$ contains two copies of identical b subunits. Thus, in plant cells if the nuclear encoded ATPase subunits have all arisen by transfer of genes from the chloroplast to the nucleus, the subunit II of CF$_0$ should be another b-type like subunit I. The structure of CF$_0$ could be envisaged as a central core of one subunit each of I, II and IV surrounded by five or more III subunits (Cox et al., 1986). Of some interest is the question of why the cyanobacterial or chloroplast F$_0$ should have a heterodimer rather than a homodimer of b-type subunits involved in interaction with the F$_1$.

CONCLUSIONS

It seems likely that the CF$_0$ sector of the chloroplast ATP synthase is composed of four subunits (Table I). The genes for three subunits are located and transcribed with the gene for the CF_1 alpha subunit on the chloroplast genome, while the fourth subunit is nuclear-encoded. From model building and studies on homologous subunits of the *E. coli* and mitochondrial ATPases, the proton pore of CF_0 is postulated to arise from interaction between subunits III and IV.

REFERENCES

Bird, C.R., Koller, B., Auffret, A.D., Huttly, A.K., Howe, C.J., Dyer, T.A. and Gray, J.C. (1985) EMBO J. 4, 1381-1388.

Cain, B.D. and Simoni, R.D. (1986) J. Biol. Chem. 261, 10043-10050.

Cox, G.B., Fimmel, A.L., Gibson, F. and Hatch, L. (1986) Biochim. Biophys. Acta 849, 62-69.

Cozens, A.L. and Walker, J.E. (1987) J. Mol. Biol. 194, 359-383.

Cozens, A.L., Walker, J.E., Phillips, A.L., Huttly, A.K. and Gray, J.C. (1986) EMBO J. 5, 217-222.

Fimmel, A.L., Jans, D.A., Langman, L., James, L.B., Ash, G.R., Downie, J.A., Senior, A.E., Gibson, F. and Cox, G.B. (1983) Biochem. J. 213, 451-458.

Foster, D.L. and Fillingame, R.H. (1982) J. Biol. Chem. 257, 2009-2015.

Hennig, J. and Herrmann, R.G. (1986) Mol. Gen. Genet. 203, 117-128.

Hudson, G.S., Mason, J.G., Holton, T.A., Koller, B.,. Cox, G.B., Whitfeld, P.R. and Bot tomley, W. (1987) J. Mol. Biol., in press.

Perlin, D.S., Cox, D.N. and Senior, A.E. (1983) J. Biol. Chem. 258, 9793-9800.

Pick, U. and Racker, E. (1979) J. Biol. Chem. 254, 2793-2799.

Sebald, W. and Hoppe, J. (1981) Curr. Top. Bioenerg. 12, 2-64.

Sebald, W. and Wachter, E. (1980) FEBS Lett. 122, 307-311.

Suss, K.-H. and Schmidt, O. (1982) FEBS Lett. 144, 213-218.

Walker, J.E., Saraste, M. and Gay, N.J. (1984) Biochim. Biophys. Acta 768, 164-200.

Watanabe, A. and Price, C.A. (1982) Proc. Natl. Acad. Sci. USA 79, 6304-6308.

Westhoff, P., Alt, J., Nelson, N. and Herrmann, R.G. (1985) Mol. Gen. Genet. 199, 290-299.

Zurawski, G., Bottomley,W. and Whitfeld, P.R. (1982) Proc. Natl. Acad. Sci. USA 79, 6260-6264.

EXPRESSION AND EVOLUTION OF THE CHLOROPLAST

ATP SYNTHASE GENES

Graham S. Hudson, John G. Mason, Barbara Koller*, Paul R. Whitfeld
and Warwick Bottomley

CSIRO, Division of Plant Industry, Canberra, ACT, Australia
and *EMBL, Heidelberg, Federal Republic of Germany

Although the majority of chloroplast polypeptides are imported from the cytosol, a number of them are transcribed and translated within the organelle (Ellis, 1981). Most of these chloroplast-encoded polypeptides are subunits of large complexes which contain at least one nuclear-encoded subunit, so that coordinated expression of chloroplast and nuclear genes is essential for correct biogenesis of the complexes. This is the case for the major thylakoid membrane complexes involved in photosynthesis; for example the H^+-ATP synthase is composed of six chloroplast- and probably three nuclear-encoded subunits (see the preceding paper). In this paper we examine the organization, expression and evolution of the chloroplast *atp* genes as an initial approach to an understanding of these processes.

STRUCTURE OF THE CHLOROPLAST GENOME

The genome from most plant chloroplasts is a closed, circular DNA of about 120 to 160 kbp (Palmer, 1985). Each chloroplast contains many, apparently identical copies of the genome, and since there may be more than 100 chloroplasts per leaf cell, the gene copy number may be as high as 5000 per cell (Scott and Possingham, 1980). A feature of the chloroplast genomes from most higher plants, except for some legumes, is the presence of two large inverted repeats which divide the DNA molecule into small and large single copy regions (Fig. 1). These repeats apparently confer stability on the genome structure, and contain the rRNA operons. The complete DNA sequence of two chloroplast genomes, from tobacco and liverwort, has been determined and shown to contain about 120 genes for tRNAs, rRNAs and proteins involved in either transcription, translation or photosynthesis (Ohyama et al., 1986;

Shinozaki et al., 1986). The six genes of the subunits of the ATP synthase are found at two loci, *atp*BE and *atp*IHFA, separated by 50 kbp in the large single copy region of the spinach chloroplast genome (Fig. 1).

THE BETA OPERON

The *atp*BE or beta operon is located approximately 700 bp upstream from, but on the opposite strand to, the gene encoding the large subunit of ribulose bisphosphate carboxylase. It encodes the beta and epsilon subunits of CF_1, and in most higher plants the two coding regions overlap by a single base such that the UGA stop codon of *atp*B is preceded by an A to form the initiator codon of *atp*E (Zurawski et al., 1982; Krebbers et al., 1982). In pea, a duplication of a 5 bp sequence has resulted in the creation of a new stop codon 22 bp upstream from the *atp*E initiator codon so that the genes do not overlap (Zurawski et al., 1986). This observation and the fact that one epsilon and three beta subunits are required per CF_1 complex suggest that translation of the two genes is not coupled. All RNA mapping data point to transcription of these

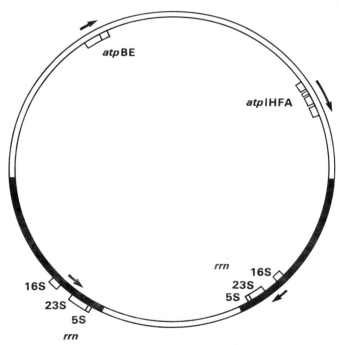

Fig 1. Map of the Spinach Chloroplast Genome Showing the Position of the Genes for ATP Synthase Subunits (*atp*BE and *atp*IHFA). Also shown are the rRNA operons (*rrn*) situated within the large, inverted repeats (shaded).

genes initiating at a single promoter several hundred bases upstream from *atp*B and translation occurring on these dicistronic mRNAs (Fig. 2A). The promoters for the spinach, pea and maize operons have been mapped to show the presence of promoter elements similar to those found upstream from *E. coli* genes as well as other chloroplast genes (Bradley and Gatenby, 1985; Gruissem and Zurawski, 1985; Mullet et al., 1985). The spinach operon has at least four mRNAs (Mullet et al., 1985), possibly due to processing at the 5'-end of the primary transcript.

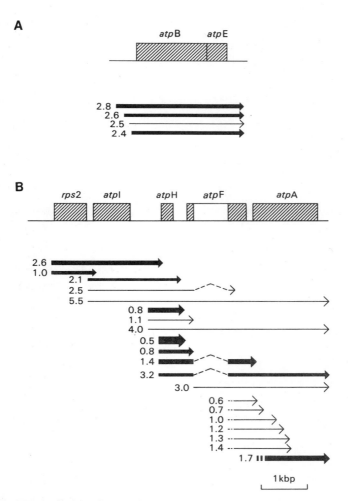

Fig. 2. Transcription of the Spinach Chloroplast *atp* Genes. (A) *atp*BE (after Zurawski et al., 1982; Mullet et al., 1985), (B) *atp*IHFA (Hudson et al., 1987). The numbers give the approximate size of the transcripts in kilobases, and the thickness of the lines reflects the relative abundance of each of the transcript sizes.

29

THE ALPHA OPERON

The *atp*IHFA or alpha operon is situated distal to the genes for three RNA polymerase subunits and one ribosomal protein, S2. The significance, if any, of this association of *rpo*BC and, *rps*2 to the *atp* genes is not yet understood. The intergenic regions range in size from 69 to 696 bp in spinach (Hudson et al., 1987). The *atp*F gene is split by a group II intron and transcripts crossing this gene are a mixture of unspliced, spliced and truncated molecules. RNA mapping has revealed a complex pattern of mono-, di- and polycistronic transcripts spanning the *rps*2 and *atp* genes, with 5'-ends upstream of *rps*2, *atp*I and *atp*H (Fig. 2B). No obvious promoter sequences are associated with the mapped 5'-ends so that it is unclear if the variety of mRNAs reflect multiple transcript initiation and/or processing events. The significant levels of unspliced and short *atp*F and *atp*A RNAs may indicate the synthesis of truncated polypeptides, a phenomenon which has been observed by *in vitro* translation of *atp*F selected mRNA (Westhoff et al., 1985). The most abundant transcripts are those of *atp*H, consistent with CF_oIII being the most abundant subunit.

EVOLUTION OF THE ATP SYNTHASE SUBUNITS

Comparison of the F_1F_o subunits and their genes from a number of bacterial, chloroplast and mitochondrial sources has allowed several observations to be made. First, there is considerable variation in the conservation of different subunits across the spectrum of F_1F_o ATP synthases (Table I). The strong conservation of the alpha and beta subunits highlights their central roles in the ATPase function of the complex, while the poor conservation of the *a* and *b* subunits may be due to lower constraints on their evolution and/or adaptation of the F_o to different membrane environments. Subunit *c* is the most highly conserved subunit amongst the chloroplasts but it is only moderately conserved between chloroplasts, mitochondria and *E. coli*. A major difference between F_1F_o complexes from different sources may lie in the acquisition of extra subunits, particularly in the case of the mitochondrial complex which has at least 11 subunits compared to the 8 of *E. coli*. The cyanobacterial and possibly the chloroplast complexes have 9 subunits, due to duplication of the *b* subunit gene (Cozens and Walker, 1987).

Second, examination of the chloroplast genes reveals a slow rate of evolution of the coding sequences. Comparing the spinach and liverwort *atp* genes and assuming that vascular and nonvascular plants separated some 430 million years ago (Raven et al., 1986), the rate of evolution of the coding sequences can be calculated as approx. 2 x 10^{-9} nucleotide substitutions per site per year which is comparable to calculated rates

Table I. Comparison of Amino Acid Sequences of F_1F_0 Subunits Relative to Those of the Spinach Chloroplast Complex.

Organism	% Amino Acid Sequence Identity					
	alpha	beta	epsilon	a(IV)	b(I)	c(III)
Tobacco chloroplast[a]	95	95	87	96	91	99
Pea chloroplast[b]	91	93	81	90	83	99
Liverwort chloroplast[c]	86	89	60	81	51	98
Cyanobacterium[d]	74	81	43	70	26	86
E. coli[e]	54	66	26	12	19	29
Bovine mitochondrion[f]	60	68		10		33

[a]Shinozaki et al. (1986). [b]Zurawski et al. (1986); Hudson et al. (1987).
[c]Ohyama et al. (1986). [d]Cozens and Walker (1987).
[e]Walker et al. (1984). [f]Anderson et al. (1982); Walker et al. (1985).

of evolution of mammalian nuclear genes. Noncoding sequences, apart from transcription, translation and processing signals, vary considerably between species, principally due to the insertion or deletion of nucleotides (see Zurawski et al., 1984; Hudson et al., 1987).

Third, there is conservation of the order of *atp* genes in plant chloroplasts relative to that in the *unc* operon of *E. coli*. The latter encodes (in order) subunits *a*, *c*, *b*, delta, alpha, gamma, beta, epsilon while the chloroplast alpha and beta operons encode IV (equivalent to subunit *a*), III (*c*), I (*b*), alpha and beta, epsilon, respectively. It is assumed that the missing *atp* genes were transferred to the nucleus. In the green alga, *Chlamydomonas reinhardtii*, the arrangement of the chloroplast *atp* genes is quite different (Woessner et al., 1987), indicating considerable scrambling of this genome.

Fourth, sequencing of the *atp* operons in purple non-sulphur bacteria and cyanobacteria has provided strong evidence for these organisms being related to the ancestors of the mitochondria and chloroplasts (Cozens and Walker, 1987). Nakamura and coworkers elsewhere in this book point out that the plant mitochondrial beta subunit is more closely related to the mammalian mitochondrial beta subunit than the chloroplast beta subunit, showing independent acquisition of the genes for mitochondrial and chloroplast ATPase subunits by the eukaryotic ancestors of plant cells.

CONCLUSIONS

The six chloroplast genes for ATP synthase subunits are organized as two operons in plants but not in the green alga, *Chlamydomonas reinhardtii*. The beta operon shows a relatively simple organization and transcription pattern, while the alpha operon is transcribed into a large variety of mRNAs. The *atp* genes are related to those of cyanobacteria and, more distantly, to those of non-photosynthetic bacteria suggesting a common origin of most F_1F_0 ATP synthase subunits.

REFERENCES

Anderson, S., de Bruijn, M.H.L., Coulson, A.R., Eperon, I.C., Sanger, F. and Young, I.G. (1982) J. Mol. Biol. 156, 683-717.

Bradley, D. and Gatenby, A.A. (1985) EMBO J. 4, 3641-3648.

Cozens, A.L. and Walker, J.E. (1987) J. Mol. Biol. 194, 359-383.

Ellis, R.J. (1981) Ann. Rev. Plant Physiol. 32, 111-137.

Gruissem, W. and Zurawski, G. (1985) EMBO J. 4, 3375-3383.

Hudson, G.S., Mason, J.G., Holton, T.A., Koller, B., Cox. G.B., Whitfeld, P.R. and Bottomley, W. (1987) J. Mol. Biol., in press.

Krebbers, E.T., Larrinua, I.M., McIntosh, L. and Bogorad, L. (1982) Nucleic Acids Res. 10, 4985-5002.

Mullet, J.E., Orozoco, E.M. Jr. and Chua, N.-H. (1985) Plant Mol. Biol. 4, 39-54.

Ohyama, K., Fukuzawa, H., Kohchi, T., Shirai, H., Sano, T., Sano, S., Umesono, K., Shiki, Y., Takeuchi, M., Chang, Z., Aota, S., Inokuchi, H. and Ozeki, H. (1986) Nature 322, 572-574.

Palmer, J.D. (1985) Ann. Rev. Genet. 19, 325-354.

Raven, P.H., Evert, R.F. and Eichhorn, S.E. (1986) in "Biology of Plants", 4th edition, Worth Publishers Inc., New York.

Scott, N.S. and Possingham, J.V. (1980) J. Exp. Bot. 31, 1081-1092.

Shinozaki, K., Ohme, M., Tanaka, M., Wakasugi, T., Hayashida, N., Matsubayashi, T., Zaita, N., Chunwongse, J., Obokata, J., Yamaguchi-Shinozaki, K.., Otoh, C., Torazawa, K., Meng,B.Y., Sugita, M., Deno, H., Kamogashira, T., Yamada, K., Kusuda, J., Takaiwa, F., Kato, A., Tohdoh, N., Shimada, H. and Sugiura, M. (1986) EMBO J. 5, 2043-2049.

Walker, J.E., Fearnley, I.M., Gay, N.J., Gibson, B.W., Northrop, F.D., Powell, J.J., Runswick, M.J., Saraste, M. and Tybulewicz, V.L.J. (1985) J. Mol. Biol. 184, 677-701.

Walker, J.E., Saraste, M. and Gay, N.J. (1984) Biochim. Biophys. Acta. 768, 164-200.

Westhoff, P., Alt, J.,. Nelson, N. and Herrmann, R.G. (1985). Mol. Gen. Genet. 199, 290-299.

Woessner, J.P., Gillham, N.W. and Boynton, J.E. (1987) Plant Mol. Biol. 8, 151-158.

Zurawski, G., Bottomley, W. and Whitfeld, P.R. (1982) Proc. Natl. Acad. Sci. USA 79, 6260-6264.

Zurawski, G., Bottomley, W. and Whitfeld, P.R. (1986) Nucleic Acids Res. 14, 3974.

Zurawski, G., Clegg, M.T. and Brown, A.H.D. (1984) Genetics 106, 735-749.

STRUCTURE AND EXPRESSION OF GENES ENCODING

HIGHER PLANT MITOCHONDRIAL F_1F_0-ATPase SUBUNITS

Kenzo Nakamura, Atsushi Morikami, Kazuto Kobayashi,
Yukimoto Iwasaki and Tadashi Asahi

Laboratory of Biochemistry, School of Agriculture
Nagoya University, Chikusa, Nagoya 464, Japan

INTRODUCTION

In higher plants, both mitochondria and chloroplasts have their own F_1F_0-ATPase, and the role of these two complexes in the supply of cellular ATP varies greatly among differentiated plant tissues. Plant mitochondrial F_1 either consist of five subunits like those from the other organisms, or contain an additional sixth subunit depending on plant species, and these F_1 subunits are distinct from subunits of the chloroplast CF_1 of the same plant at least in several plant species examined. Unlike fungal and mammalian F_1, where all the subunits are encoded by nuclear genes, the α-subunit of plant F_1 is encoded by the mitochondrial genome and all the other subunits seem to be encoded by the nuclear genome. Detailed studies on plant mitochondrial F_1 subunits and their genes, and their comparison with chloroplastic counterparts, are expected to provide valuable informations on structure, biosynthesis, regulation and evolution of two F_1 complexes in plant cells.

PURIFICATION AND CHARACTERIZATION OF PLANT MITOCHONDRIAL F_1-ATPase SUBUNITS

Subunit composition of plant mitochondrial F_1 purified from several sources are summarized in Table I. Unlike F_1 from the other organisms, the sweet potato root F_1 consists of six subunits, and the additional sixth subunit has been designated as δ'-subunit (Iwasaki and Asahi, 1983). The pea cotyledon F_1 preparation by Horak and Packer (1985) shows subunit composition and subunit molecular weight similar to those of the sweet potato F_1. While the bean F_1 purified by Boutry et al. (1983) was reported to contain only five subunits, the molecular weight of its ϵ-subunit was closer

Table I. Subunits of Higher Plant Mitochondrial F_1-ATPase and Chloroplast CF_1-ATPase.

Source	Subunit (kDa)						Reference
	α	β	γ	δ	δ'	ϵ	
Sweet potato	52.5	51.5	35.5	27	2312		Iwasaki et al. (1983)
Pea	57	55	36.5	26.5	25	8	Horak & Packer (1985)
Bean	52	51	34	23.8	? ←	22.9	Boutry et al. (1983)
Maize	58	55	35	22		12	Spitsberg et al. (1985)
Oat	58	55	35	22		14	Randall et al. (1985)
Pea (CF_1)	54	54	39	22.5		13	Horak & Packer (1985)
Maize (CF_1)	60	56	40	22.5		15.5	Spitsberg et al. (1985)

to that of the δ'-subunits from sweet potato and pea. In contrast to these F_1 complexes from dicotyledonous plants, F_1 complexes purified from two monocotyledonous plants, maize (Spitsberg et al., 1985) and oat (Randall et al., 1985), apparently contain only five subunits.

The subunits of sweet potato mitochondrial F_1 were purified from the complex by SDS-polyacrylamide gel electrophoresis or by gel filtration and ion-exchange high-performance liquid chromatography (Iwasaki and Asahi, 1985; Kobayashi et al., 1986). Tryptic peptide maps of the purified δ- and δ'-subunits were quite different from each other. However, precursors for δ- and δ'-subunits synthesized *in vitro* by poly(A)$^+$RNA from sweet potato root were immunologically indistinguishable from each other. These results suggest that δ- and δ'-subunits are different but immunologically related polypeptides.

The determination of the N-terminal 18 amino acid sequence of the purified β-subunit showed that there are equimolar amounts of Glu and Asp at the eight position (Kobayashi et al., 1986) (Fig. 1). At least two precursors for the β-subunit which is 6,500-7,500 daltons larger than the mature form, and multiple precursors for δ- and δ'-subunits, were detected by the *in vitro* translation studies (Iwasaki and Asahi, 1985). These results suggest that the β-subunit of sweet potato F_1 contains at least two major polypeptides similar to each other. Boutry and Chua (1985) have reported that tobacco F_1 β-subunit is encoded by two nuclear genes, both of which are expressed.

The N-terminal amino acid sequence of the β-subunit showed homology to amino acid sequence of the tobacco F_1 β-subunit precursor deduced from the nucleotide sequence (Boutry and Chua, 1985) between residues 56 and 73, suggesting that the N-terminal 55 amino acids of the tobacco precursor constitute the presequence required for mitochondrial targeting. The putative presequence for tobacco F_1 β-subunit does not show any sequence homology with that of human F_1 β-subunit reported by Ohta and Kagawa (1986). It shows, however, structural characteristics of presequences of many imported mitochondrial proteins from animal and fungi; it is poor in acidic amino acids and enriched in basic amino acids, serine and threonine, and it can form amphiphilic helical structure (von Heijne, 1986). The transit peptides of nuclear-encoded chloroplast proteins are also rich in basic amino acids, serine and threonine. The putative presequence for tobacco F_1 β-subunit, however, does not show homology to any one of the framework sequences shared by many transit peptides (Karlin-Neumann and Tobin, 1986).

F_1-ATPase α-SUBUNIT GENE LOCATE AT THE HOMOLOGOUS RECOMBINATION SITE OF THE PEA MITOCHONDRIAL GENOME

The mitochondrial genomes of higher plants are much larger and more complex than those in animal and fungal cells, varying from 218 kb in *Brassica* to an estimated 2400 kb in muskmelon. Furthermore, recent studies indicate that several plant mitochondrial genomes are carried on discrete circular molecules which are derived from each other by homologous recombinations between repeat sequences present in "master circle" (for reviews, Stern and Palmer, 1984a; Falconet et al., 1984).

The F_1 α-subunit coding sequence is repeated in the pea mitochondrial genome (Morikami and Nakamura, 1987a). Restriction enzyme mapping analyses of 32 partial *Sau*3A fragments of mitochondrial DNA independently cloned into λ phage vector indicated that these cloned segments can be classified into four types according to their structures (Fig. 2). These four types share the core repeat region of 1.7 kb flanked by four paired combinations of two left and two right unique sequences. This pattern of fragments could be a typical result of a homologous recombination event. Homologous recombination between two of these repeat sequences, either [types I and

```
                          Glu
  -X- -X- Ala Ala Ala Pro Ala    Lys Pro Ala Ala Lys Pro Ala -X- Asn Glu
                          Asp
  1             5              10                 15
```

Fig. 1. N-terminal Amino Acid Sequence of Purified β-Subunit.

III] or [types II and IV], can generate the other two types. Probably, one of these paired sequences are present on a 430 kb master circular DNA of the pea mitochondrial genome.

Nucleotide sequencing of types II and IV alleles indicated that the uninterrupted α-subunit coding sequence of 1521 bp is present in two of these loci. A rearrangement 965 bp 3' to the ATG initiation codon generates two copies of pseudogenes where the C-terminal two-thirds of the α-subunit coding sequence is replaced with unidentified coding frame. The other border of sequence divergence is located 733 bp upstream of the ATG initiation codon.

Recombination near the α-subunit gene has also been reported in fertile maize (Isaac et al., 1985) and *Oenothera* (Schuster and Brennicke, 1986). However, the structure of repeats involved in the recombination varies for each mitochondrial genome. In the case of fertile maize, the entire α-subunit coding sequence is located near the end of a 12 kb sequence directly repeated at two locations in the master circle. On the other hand, both of the left- and right- hand borders of a 264 bp recombination repeats are located within the *Oenothera* α-subunit coding sequence. Obviously, one functional α-subunit gene on a master mitochondrial genome is enough for the vital function of mitochondrial F_1-ATPase.

STRUCTURE OF PEA MITOCHONDRIAL F_1-ATPase α-SUBUNIT GENE

The nucleotide sequence of the pea F_1 α-subunit gene shares 92-94% homologies with α-subunit genes of maize (Isaac et al., 1985; Braun and Levings III, 1985) (Fig. 3a) and *Oenothera* (Schuster and Brennicke, 1986) mitochondria.

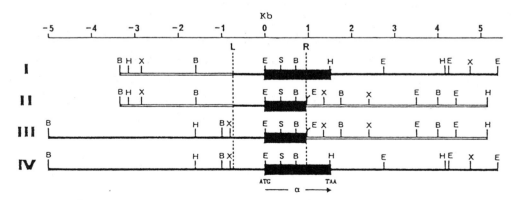

Fig. 2. Restriction Enzyme Mapping Analyses of Partial *Sau*3A Fragments of Mitochondrial DNA Independently Cloned into λ Phage Vector.

Homologies at the amino acid sequence level are about 95% among these three F_1 α-subunits (Fig. 4). About half of these amino acid replacements are confined within the C-terminal 50 amino acid residues.

The pea CF_1 α-subunit gene encoded by the chloroplast genome (P.R. Whitfeld, personal communication) shows only limited nucleotide sequence homology with that of the pea mitochondrial F_1 α-subunit gene (Fig. 3b), and the amino acid sequence of these two pea α-subunits share only 60% homology (Fig. 4). Since CF_1 α-subunit gene is highly conserved among several plant species (Fig. 3c, Fig. 4), these results seem to support the idea that the mitochondrial and chloroplast genes for F_1α and CF_1α originate from different symbiotic organisms and evolved independent pathways in spite of the fact that some genetic exchange is occurring between these two organelles (Stern and Palmer, 1984b).

EXPRESSION OF PEA MITOCHONDRIAL F_1-ATPase α-SUBUNIT GENE AND ITS PSEUDOGENE

Single-strand probes derived from the α-subunit coding sequence hybridized strongly to three mitochondrial RNA bands of approximately 2.7, 2.3 and 1.8 kb, and weakly to a 4.5 kb band and several other minor bands. Among these, a 2.3 kb band was the most abundant. Preliminary mapping of these α-subunit gene transcripts suggest that the 5'-termini of the three major transcripts are localized within a 456 bp sequence upstream from the ATG initiation codon, and the 5'-terminus of the minor 4.5 kb transcript is localized more than 1.6 kb upstream from the ATG codon (Fig. 5). However, since it is not known whether these 5'-termini of the transcripts represent

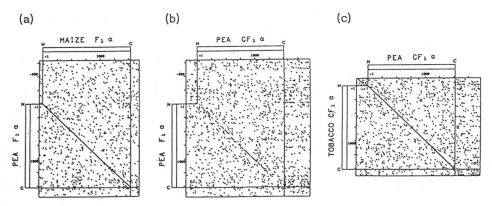

Fig. 3. Homologies Between the Nucleotide Sequences of the Pea and Maize Mitochondrial and Chloroplast F_1 α–Subunit Genes.

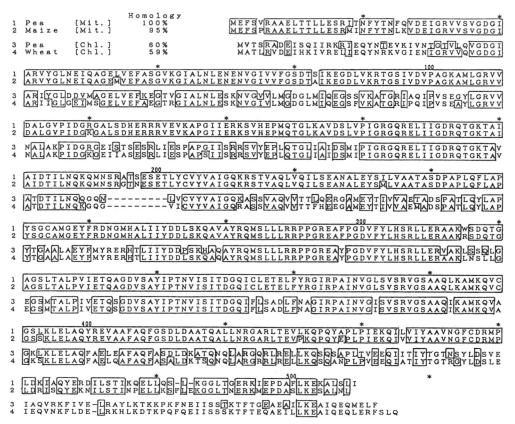

Fig. 4. Homologies Between the Amino Acid Sequences of the Pea and Maize Mitochondrial and Chloroplast F_1 α-Subunits.

true transcription initiation sites or the processing sites of the larger primary transcript, it can not be determined whether both of the α-subunit genes in types I and IV alleles are functional or not.

No transcript which is contiguous from the α-subunit coding sequence to the right-hand unique sequence in the types II and III alleles is detected (Fig. 5), and no truncated α-subunit-related polypeptide is detected among the *in vitro* translation products by isolated pea mitochondria. These results suggest that the α-subunit coding sequence transcripts in types II and III alleles are very rapidly degraded and the α-subunit pseudogenes in types II and III alleles do not express any detectable amount of unusual α-subunit-like polypeptide.

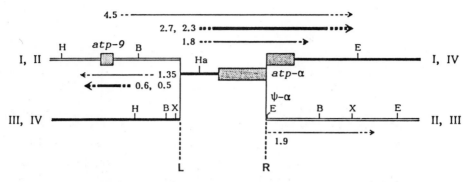

Fig. 5. Expression of Pea Mitochondrial F_1-ATPase α-Subunit Gene.

F_0-ATPase SUBUNIT 9 GENE LOCATE UPSTREAM OF F_1-ATPase α-SUBUNIT GENE IN THE PEA MITOCHONDRIAL GENOME

During the course of transcript mapping of the α-subunit gene(s), we found the presence of a strong transcription unit in one of the two sequences upstream of the 1.7 kb repeat sequence (Fig. 5). Nucleotide sequencing indicated that this transcription unit codes for the F_0-ATPase proteolipid or subunit 9 (Fig. 6). The gene starts 2.1 kb upstream of the α-subunit initiation codon, and is transcribed in the opposite direction with respect to the α-subunit gene (Morikami and Nakamura, 1987b).

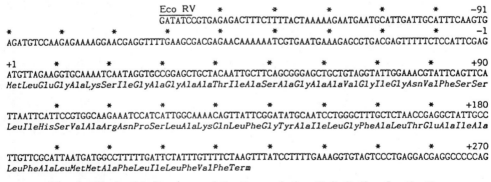

Fig. 6. Nucleotide Sequence of the Transcription Unit Coding for the F_0-ATPase Proteolipid or Subunit 9 of the Pea Mitochondrial F_0-ATPase.

ACKNOWLEDGMENT

We are grateful to Dr. P.R. Whitfeld for providing pea CF_1 α-subunit gene sequence before publication.

REFERENCES

Boutry, M., Briquet, M. and Goffeau, A. (1983) J. Biol. Chem. 258, 8524-8526.

Boutry, M. and Chua, N.H. (1985) EMBO J. 4, 2159-2165.

Braun, C.J. and Levings III, C.S. (1985) Plant Physiol. 79, 571-577.

Deno, H., Shinozaki, K. and Sugiura, M. (1983) Nucleic Acids Res. 11, 2185-2191.

Falconet, D., Lejeune, B., Quetier, F. and Gray, M.W. (1984) EMBO J. 3, 297-302.

Hack, E. and Leaver, C.J. (1983) EMBO J. 2, 1783-1789.

Horak, A. and Packer, M. (1985) Biochim. Biophys. Acta 810, 310-318.

Isaac, P.G., Brennicke, A., Dunbar, S.M. and Leaver, C.J. (1985) Curr. Genet. 10, 321-328.

Iwasaki, Y. and Asahi, T. (1983) Arch. Biochem. Biophys. 227, 164-173.

Iwasaki, Y. and Asahi, T. (1985) Plant Mol. Biol. 5, 339-346.

Iwasaki, Y., Matsuoka, M. and Asahi, T. (1984) FEBS Lett. 171, 249-252.

Karlin-Neuman, G.A. and Tobin, E.M. (1986) EMBO J. 5, 9-13.

Kobayashi, K., Iwasaki, Y., Sasaki, T., Nakamura, K. and Asahi, T. (1986) FEBS Letters 203, 144-148.

Morikami, A. and Nakamura, K. (1987a) J. Biochem. (Tokyo) 101, 967-976.

Morikami, A. and Nakamura, K. (1987b) Nucleic acids Res., in press.

Ohta, S. and Kagawa, Y. (1986) J. Biochem. (Tokyo) 99, 135-141.

Randall, S.K., Wang, Y. and Sze, H. (1985) Plant Physiol. 79, 957-962.

Schuster, W. and Brennicke, A. (1986) Mol. Gen. Genet. 204, 29-35.

Spitsberg, V.L., Pfeiffer, N.E., Partridge, B., Wylie, D.E. and Schuster, S.M. (1985) Plant Physiol. 77, 339-345.

Stern, D.B. and Palmer, J.D. (1984a) Nucleic Acids Res. 12, 6141-6157.

Stern, D.B. and Palmer, J.D. (1984b) Proc. Natl. Acad. Sci. USA 81, 1946-1950.

von Heijne, G. (1986) EMBO J. 5, 1335-1342.

POLYPEPTIDE SYNTHESIS, IMPORT AND ASSEMBLY

THE ASSEMBLY OF THE F_1F_0-ATPase COMPLEX IN *ESCHERICHIA COLI*

Frank Gibson and Graeme B. Cox

Biochemistry Department, John Curtin School of Medical Research
Australian National University, Canberra, Australia

Two possible routes for the assembly of the F_1F_0-ATPase are (a) the separate formation of the F_0 and F_1-ATPase and their subsequent assembly (the **separate sector assembly pathway**) and (b) a pathway in which the assembly does not proceed as above, but involves an interaction between various subunits of the F_0 and F_1 before the individual sectors are completed (the **integrated membrane assembly pathway**). Studies, on *Escherichia coli*, which bear on this problem have been carried out in a number of laboratories.

The assembly *in vitro* of *E. coli* F_1-ATPase was first reported by Dunn and Futai (1980). The assembled F_1-ATPase reconstituted oxidative phosphorylation in F_1-depleted membranes. Purified F_0 subunits from *E. coli* were later assembled in phospholipid vesicles by Schneider and Altendorf (1985) and would then bind F_1-ATPase and translocate protons. These results support the model for the assembly of the F_1F_0-ATPase complex proposed by Sternweis and Smith (1977) designated above as the **separate sector assembly pathway**.

That the F_1 and F_0 sectors of the F_1F_0-ATPase might not be assembled separately *in vivo* was indicated by the isolation of a strain carrying a polar mutation affecting the formation of the β- and ϵ-subunits of the F_1-ATPase (Cox et al., 1981). Such strains were unable to integrate the b-subunit, encoded by the *uncF* gene, into a functional F_0 even though the genes encoding the F_0 subunits were not affected by the mutation. Further experiments, using polarity mutants carrying different plasmids, showed that the α- and β-subunits, but not the ϵ- or γ-subunits, were apparently required for the formation of a functional F_0 (Cox et al, 1981). Similar results were obtained by Friedl et al. (1983) with a polarity mutant in which the genes coding for the F_0 subunits were normal but which was unable to form the α-, β- and γ-subunits of

Molecular Structure, Function, and Assembly of the ATP Synthases
Edited by Sangkot Marzuki
Plenum Press, New York

the F_1-ATPase. It was demonstrated by immunoblotting that, while the c-subunit was present in normal amounts in the membranes of the normal strain, the b-subunit was absent.

There are a number of mutant strains in which incomplete F_1 assemblies appear to be present in membrane preparations (Fayle et al, 1987; Cox et al, 1981; Noumi and Kanazawa, 1983). The examination of two-dimensional electrophoretograms in which the apparent relative proportions of the F_1 subunits were examined, allowed the following pathway to be postulated (Cox et al., 1981) :

$$ac \rightarrow \beta ac_n \rightarrow \alpha\beta ac_n \rightarrow \alpha\beta ab_2 c_n \rightarrow \alpha\beta_2 ab_2 c_n \rightarrow \rightarrow \rightarrow \alpha_3\beta_3\gamma\delta\epsilon ab_2 c_n$$

Mutations in the $uncF$ gene coding for the b-subunit have been shown to affect assembly (Jans et al., 1984; Porter et al., 1985). For example, a haploid strain carrying a Gly-9 to Asp substitution does not insert a b-subunit into the membrane unless the mutant allele is carried on a multicopy plasmid, in which case the assembly complex so formed is functional (Jans et al, 1984). A partial revertant strain isolated from the haploid strain carrying the Gly to Asp change retained this substitution and genetic complementation tests indicated that the compensating change had occurred in the $uncA$ gene coding for the α-subunit. Such a result would support the concept of an interaction between the b-subunit and the α-subunit during assembly. It is of interest that the Gly-9 on the b-subunit would be expected to be imbedded in the membrane (Senior, 1983; Walker et al., 1984) and unable to interact with the α-subunit after the insertion of the b-subunit into the membrane.

Experiments (Friedl et al., 1983; Klionsky et al., 1983; Aris et al., 1985; Fillingame et al., 1986; von Meyenburg et al., 1985) using genetically engineered strains of $E.$ $coli$ carrying genes encoding the F_0 subunits on plasmids and chromosomal deletions have given results which indicate that in such strains it is possible to obtain a proton pore in the absence of F_1 subunits. In some of these experiments the results are difficult to interpret because of the presence of chromosomal genes coding for F_1 subunits (see Cox and Gibson, in press). Furthermore, the rapid increase in synthesis of subunit-a alone, using a heat-inducible plasmid carrying the $uncB$ gene in a strain in which the genes coding for the F_0 subunits had been deleted, allows proton conduction across the membranes (von Meyenburg et al., 1985). It was suggested that this phenomenon is due to the formation of subunit-a multimers before integration into the membranes. It would

Table I.   ATP-dependent Atebrin Fluorescence-Quenching Activities of Membranes:- Before and After Reconstitution of Stripped Membrane with F_1-ATPase.

Strain	Mutation(s)[1]	ATP-dependent At-Fluor-Q of membranes (%)	
		Native	Reconstituted[2]
Wild type	-	85	85
AN1518	Gly-48 to Asp	<5	20
AN2540	Gly-48 to Asp Pro-47 to Ser	50	20
AN2618	Gly-48 to Asp Pro-47 to Thr	50	20

[1]In the *uncC* gene coding for the ϵ-subunit.   [2]Membranes stripped and then reconstituted with saturating levels of F_1.

seem, therefore, that the experimental results obtained with strains carrying the F_o genes on multicopy plasmids does not reflect the situation in haploid cells.

Genetically engineered strains have also been used to examine the assembly of the F_1-ATPase from plasmids carrying the genes coding for the F_1-subunits in strains lacking the genes coding for the F_o-subunits (Klionsky and Simoni, 1985). Evidence was obtained for the formation of fully functional F_1-ATPase *in vivo* in such strains. However, the examination of subunit aggregates precipitated by subunit specific antibodies indicated that a number of such aggregates were present in the cytoplasm. The aggregate corresponding to the normal F_1-ATPase appeared to be only a minor component. This is surprising if the normal pathway of assembly of the F_1-ATPase is in the cytoplasm.

A mutant strain of *E. coli* (AN1518) unable to carry out oxidative phosphorylation has recently been isolated in which the Gly-48 of the ϵ-subunit has been replaced by Asp (Cox et al., 1987). Two partial revertants, which could carry out oxidative phosphorylation were isolated from strain AN1518. In both of these strains the original Gly to Asp change was still present, but the adjacent Pro-47 had been replaced by Ser in one strain (AN2540) and by Thr in the other strain (AN2618). A comparison of the properties of the F_1-ATPase from strain AN1518 and two partial revertants isolated from it are difficult to reconcile with the assembly of a preformed

F_1F_0-ATPase onto a pre-existing F_0. Thus, as shown in Table I, the atebrin fluorescence-quenching properties of the native membranes from the mutant and the two partial revertants derived from it differ markedly. However, if the F_1-ATPase from each of these membranes is stripped off and reconstituted using stripped normal membranes, the abilities to carry out ATP-dependent atebrin fluorescence-quenching are now identical although, in each case, they differ from those of the native membranes.

It would seem from the experimental evidence that the pathway of assembly of the F_1F_0-ATPase in the normal haploid *E. coli* cell is likely to follow a pathway such as that proposed as the integrated membrane assembly pathway. A more detailed assessment of the experimental evidence relating to the assembly of the F_1F_0-ATPase is presented elsewhere (Cox and Gibson, in press).

REFERENCES

Aris, J.P., Klionsky, D.J. and Simoni, R.D. (1985) J. Biol. Chem. 260, 11207-11215

Cox, G.B. and Gibson, F. (1987) Curr. Topics Bioenerg., in press

Cox, G.B., Downie, J.A., Langman, L., Senior, A.E., Ash, G., Fayle, D.R.H. and Gibson, F. (1981) J. Bacteriol. 148, 30-42

Cox, G.B., Hatch, L., Webb, D., Fimmel, A.L., Lin, Z.-H., Senior, A.E. and Gibson, F. (1987) Biochim. Biophys. Acta 890, 195-204

Dunn, S.D. and Futai, M. (1980) J. Biol. Chem. 255, 113-118

Fayle, D.R.H., Downie, J.A., Cox, G.B., Gibson, F. and Radik, J. (1987) Biochem. J. 172, 523-531

Fillingame, R.H., Porter, B., Hermolin, J. and White, L.K. (1986) J. Bacteriol. 165, 244-251

Friedl, P., Hoppe, J., Gunsalus, R.P., Michelson, O., von Meyenburg, K. and Schairer, H.U. (1983) EMBO J. 2, 99-103

Jans, D.A., Fimmel, A.L., Hatch, L., Gibson, F. and Cox, G.B. (1984) Biochem. J. 221, 43-51

Klionsky, D.J. and Simoni, R.D. (1985) J. Biol. Chem. 260, 11200-11206

Klionsky, D.J., Brusilow, W.S. and Simoni, R.D. (1983) J. Biol. Chem. 258, 10136-10143

Noumi, T. and Kanazawa, H. (1983) Biochem. Biophys. Res. Commun. 111, 143-149

Porter, A.C., Kumamoto, C., Aldape, K. and Simoni, R.D. (1985) J. Biol. Chem. 260, 8182-8187

Schneider, E. and Altendorf, K. (1985) EMBO. J. 4, 515-518

Senior, A.E. (1983) Biochim. Biophys. Acta 726, 81-95

Sternweis, P.C. and Smith, J.B. (1977) Biochemistry 16, 4020-4025

von Meyenburg, K., Jorgensen, B.B., Michelson, O., Sorensen, L. and McCarthy, J.E.G. (1985) EMBO J. 4, 2357-2363

Walker, J.E., Saraste, M. and Gay, N.J. (1984) Biochim. Biophys. Acta 768, 164-200

BIOCHEMICAL ANALYSES OF *oli1* AND *oli2* GENE MUTATIONS

DETERMINING PRIMARY SEQUENCE CHANGES IN SUBUNITS 9 AND 6

OF YEAST ATP SYNTHASE

Sybella Meltzer, Tracy A. Willson, Linton C. Watkins, Phillip Nagley,
Sangkot Marzuki, Anthony W. Linnane and H.B. Lukins

Department of Biochemistry and Centre for Molecular Biology
and Medicine, Monash University, Clayton, Victoria 3168, Australia

SUMMARY

The biochemical properties of mitochondrial mutants having primary sequence changes in either subunit 9 or 6 of yeast mitochondrial H^+-ATPase have been examined. These *mit⁻* mutants display reduced ATPase activity which is uncoupled from proton translocation as indicated by insensitivity to inhibition by oligomycin. Several mutants with lesions in subunit 9, particularly those lacking C-terminal residues, assemble neither subunit 9 polypeptides nor subunits 8 and 6 to the H^+-ATPase complex, as determined by immunoprecipitation. With the exception of one mutant forming a longer subunit 9 polypeptide, all mutants that fail to integrate subunit 9 into the complex show a pleiotropic effect in which there is a marked reduction in the synthesis of respiratory enzyme complexes, notably in cytochrome oxidase. Thus subunit 9 not only plays a key role in the assembly of the F_o-sector of H^+-ATPase but it also has modulatory effects on the assembly of cytochrome oxidase. Mutations located in the hydrophilic loop region of subunit 9 have indicated a positive charge between residues 35 and 39 is required for H^+-ATPase function but not for the assembly of the complex.

A series of mutants having truncated subunit 6 polypeptides of varying lengths has been used to show that a polypeptide as short as 110 residues can be assembled into the H^+-ATPase complex. Subunit 6 was shown not to be required for the assembly of subunits 9 and 8 with the F_1-sector. The activity of the H^+-ATPase is affected by amino acid changes localized to two conserved transmembrane regions of subunit 6.

INTRODUCTION

Three subunits of the ATP synthase complex (H^+-ATPase) of yeast mitochondria are encoded in the mitochondrial genome and synthesized on mitochondrial ribosomes. These three mitochondrial products, subunits 6, 8 and 9, are hydrophobic proteins which comprise the membrane-integrated F_o-sector of the H^+-ATPase complex. Collectively these three subunits function in proton translocation through the membrane and in coupling of proton flux to ATP synthesis catalyzed by the F_1-sector of the complex.

For the examination of the roles of these three subunits in yeast H^+-ATPase, this laboratory utilizes mutants having mutations in either the *oli1*, *oli2* or *aap1* genes, which encode subunits 9, 6 and 8 respectively. The aim is to distinguish polypeptide domains and amino acid residues essential for proton channel function or assembly by correlating structural changes in these subunits, as determined by DNA sequence analysis of the mutant genes, with biochemical and immunological analyses of H^+-ATPase activity and assembly in the mutants. Three classes of mitochondrial mutations are employed. Mutations to oligomycin resistance in the *oli1* (Nagley et al., 1986) and *oli2* (John and Nagley, 1986) genes have indicated that the two membrane-spanning arms of subunit 9 and two conserved transmembrane regions of subunit 6 interact to form a complex oligomycin binding region in the membrane. Mutations of the *mit*⁻ class in each of the three genes have been characterized by gene sequencing (Ooi et al., 1985, for *oli1* gene; John et al., 1986, for *oli2* gene; and Macreadie et al., 1983, for *aap1* gene) and correlated with biochemical changes for a representative number of mutants (*oli1* gene, Linnane et al., 1985; Jean-Francois et al., 1986; oli2 gene, Choo et al., 1985; *aap1* gene, Macreadie et al., 1982). Lastly, intragenic revertants of *mit*⁻ mutants have provided a means for the introduction of additional amino acid substitutions into each of the three F_o subunits (subunit 6, John et al., 1986; subunit 8, Macreadie et al., 1983; subunit 9, Willson and Nagley, 1987).

This article deals with a series of correlative studies relating the biochemical properties of *mit*⁻ mutants and revertant strains with known changes in different regions of subunits 9 and 6. The biochemical changes occur not only in the assembly and/or activity of the H^+-ATPase complex, but also in other respiratory complexes of the mitochondrial inner membrane.

BIOCHEMICAL CORRELATES OF LESIONS IN SUBUNIT 9: A MEMBRANE-MODULATORY ROLE

The DCCD-binding proteolipid, subunit 9 of yeast H^+-ATPase, has for some time been considered to have an important role in the function of the proton channel of the complex (Sebald and Hoppe, 1981) and further, studies in our laboratory (Marzuki et al., 1983; Hadikusumo et al., 1984; Linnane et al., 1985) have indicated that this protein plays a key role in the assembly of the F_o-sector. This protein (76 amino acid residues) is envisaged to have a hairpin structure in the membrane in which two hydrophobic transmembrane helices are separated by a hydrophilic segment which forms a loop protruding from the matrix side of the inner membrane (Sebald and Hoppe, 1981). Distributed within each of these three segments are highly conserved amino acid residues, as shown by the analysis of proteolipids from a range of organisms (for review, see Sebald and Hoppe, 1981). This implies that these residues are essential in determining the structure and function of the proteolipid in the H^+-ATPase complex.

As a direct approach to defining functionally important structural features of subunit 9, this laboratory has studied several mit^- mutants, or revertants of such strains, having primary sequence changes located in the N-terminal, C-terminal or hydrophilic loop regions of the molecule. A comprehensive compilation of the biochemical properties of our range of $oli1\ mit^-$ mutants, including examples of several intragenic revertants is given in Table I. All the mit^- mutants retain significant, although reduced, levels of ATPase activity, being 20-60% that of wild-type (J69-1B). The oligomycin sensitivity of the residual ATPase activity is of a negligible magnitude, approaching that obtained with a mtDNA-less petite strain (EJO) devoid of all three F_o-subunits. In addition, the ATP-P_i exchange activity of each mit^- mutant is negligible (data not shown; Jean-Francois et al., 1986). These observations indicate that in the mutant strains either proton translocation through the F_o-sector is impaired or is uncoupled from F_1 activity.

Although these mit^- mutants have defective H^+-ATPase as the result of changes in subunit 9, it was of interest to observe that the mutants were suppressed also in respiratory activity. The respiratory rates were 25% or less of the wild-type value and in some mutants approached that observed with a respiratory deficient rho^o strain. Even those mutants retaining the higher values of approximately 25% residual respiration (strains 2422, 811, 15B2) did not, however, show ADP-dependent respiratory control, consistent with the uncoupled nature of the ATPase activity. The overall low rates of respiration observed in the mit^- mutants are paralleled by markedly reduced amounts of cytochromes aa_3, b and c_1 and low activities of cytochrome c oxidase and NADH-cytochrome c reductase (Jean-Francois et al., 1986). The most severely affected activity is that of cytochrome oxidase (see Table I) with two groups of mutants being distinguished, those retaining 10-20% of the wild-type activity (strains MJ1-1, 15B2, 811 and 2422) and those having about 0.5% (strains 5726, 51223, 5208, 38.6.1 and 5102). These two groupings are also reflected by their cytochrome content and NADH-cytochrome c reductase activities (data not shown).

Table I. Mitochondrial H$^+$-ATPase and Respiratory Enzyme Activities in *oli1 mit*$^-$ and Revertant Strains

Strain	*oli1* Gene Product			ATPase		Respiratory Enzyme Activities	
	Amino acid change(s)	Predicted length	Assembly	Specific activity (mmol/min/mg)	% inhibition by oligomycin	Respiration[a] (ng atom 0/min/mg)	Cytochrome oxidase (mmol/min/mg)
J69-1B (wild-type)	none	76	yes	0.97	80	128 (2.4)	1.1
EJO (rho^0)	–	–	–	0.8	13	8 (–)	0
C-terminal mutants							
5726	frameshift[b]	65	no	0.36	14	15 (–)	0.005
51223	Ser69|stop	68	no	0.36	21	16 (–)	0.006
5208	Ser69|stop	68	no	0.26	12	16 (–)	0.006
MJ1-1	frameshift[b]	84	no	0.19	32	22 (–)	0.1
N-terminal mutants							
38.6.1	frameshift[b]	7	no	0.59	21	13 (–)	0.005
5102	Gly18|Asp	76	no	0.57	20	18 (–)	0.006
15B2	Gly23|Asp	76	yes	0.21	30	24 (–)	0.08
811[c]	Gly23|Cys	76	yes	0.41	22	32 (–)	0.3
Loop region mutants							
2422	Arg39|Met	76	yes	0.24	27	33 (–)	0.2
2422-R1	Arg39|Lys	76	n.d.[d]	1.07	65	85 (2.5)	n.d.
2422-R9	Arg39|Met Asn35|Lys	76	n.d.	0.96	60	140 (1.7)	n.d.
2422-R10	Arg39|Met Asn35|Lys Leu53|Phe	76	n.d.	1.0	75	122 (1.9)	n.d.

Strains were grown in glucose-limited chemostat cultures as previously described (Marzuki and Linnane, 1979). The mutations were defined by DNA sequencing (Ooi et al., 1985; Willson et al., 1986; Willson and Nagley, 1987). Assembly of the *oli1* gene products in the H^+-ATPase complex was determined by immunoprecipitation with an anti-β monoclonal antibody (Hadikusumo et al., 1984). ATPase and respiratory activities were measured as described previously (Jean-Francois et al., 1986).

[a] Respiration determined in the presence of ethanol and ADP (Figures in brackets are respiratory control ratios).

[b] Amino acid residue changes (with first residue number) due to frameshift mutation:

5726 (57)YQKTQUYSV MJ1-1 (72)FIIRCIIYIIYYK 38.6.1 (5)FSS

[c] Strain 811 is a temperature-sensitive *mit⁻* mutant. Biochemical data obtained by growing at the non-permissive growth temperature of 18°C.

[d] n.d. not determined

An analysis of the assembly data of these mutants provides insight into these two groups. The assembly of subunit 9 to the F_1- sector and to the other F_0-subunits, 6 and 8, was studied using a monoclonal anti-β antibody which immunoprecipitates assembled H^+-ATPase subunits from Triton extracts of mitochondria (Hadikusumo et al., 1984; Jean-Francois et al., 1986). As listed in Table I, subunit 9 is not assembled in strains 5726, 51223, 5208, 38.6.1 and 5102 while this protein is immunoabsorbed to the antibody in strains 15B2, 811 and 2422. In general, those strains that have extremely low cytochrome oxidase activities, 0.5% of wild-type, do not assemble subunit 9; while those strains having cytochrome oxidase activities 10% to 20% of wild-type assemble subunit 9 into the H^+-ATPase complex together with the other F_0-sector subunits. This pleiotropic effect suggests that subunit 9 plays a key role not only in the assembly of the F_0-sector, as previously reported (Marzuki et al.; 1983; Linnane et al., 1985; Marzuki et al., this volume), but also in the general organisation of the inner mitochondrial membrane, since the absence of subunit 9 (strain 38.6.1) or the inability to assemble subunit 9 to the H^+-ATPase (strains 51223, 5208 and 5102) results in a failure to develop an active respiratory system.

One *oli1 mit⁻* mutant (MJ1-1), recently isolated in this laboratory, cannot be readily assigned to either of the above groups, but may provide new insight into the membrane-modulatory role of subunit 9. The *oli1* gene product of this mutant has the first 71 amino acid residues of the wild-type subunit 9 protein with an additional 13 residues forming a novel C-terminus (Willson et al., 1986). Initial measurements showed relatively high cytochrome aa_3 content and cytochrome oxidase activity in this mutant (see Table I). However, analysis of mitochondrial translation products showed the presence of only very small amounts of subunit 9, which in turn, could not be detected in immunoprecipitates of the H^+-ATPase complex. It is possible in this exceptional case, that subunit 9 is present in low amounts in the membrane and that the extensive residue changes at the C-terminus renders unstable the binding of this protein to the complex.

If a requirement for the functional assembly of cytochrome oxidase in the inner mitochondrial membrane is the presence therein of assembled F_0-sectors of H^+-ATPase, then one would infer that the very small quantities of the modified subunit 9 protein made in MJ1-1 do lead to assembled F_0-sectors, but at a level beyond detection by our immunoprecipitation techniques. On the other hand, the behaviour of strain MJ1-1 is consistent with the view that the modulation of general membrane assembly, represented by development of functional cytochrome oxidase activity, may be effected by subunit 9 before it assembles into the F_0-sector. Certainly in mutant 38.6.1 there is no subunit 9 gene product to enter the membrane, and it is possible that the modified subunit 9 of the other non-assembling *oli1 mit⁻* mutants such as 51223, 5208 and 5102 never enters the membrane.

The membrane-modulatory role of subunit 9 is emphasized by recent observations on two temperature-sensitive nuclear *pet* mutants of yeast. After transfer of such cells to the non-permissive temperature of 36°C the biosynthesis of subunit 9 is specifically abolished, and this is followed some hours later by severe impairments in the assembly of the cytochrome oxidase complex (M.J. Payne and H.B. Lukins, unpublished data).

ROLE OF INDIVIDUAL DOMAINS OF SUBUNIT 9

Deductions concerning amino acid residues and peptide domains important for assembly of subunit 9 can be made by relating changes in the polypeptides with the findings of immunoprecipitation experiments. Thus, all four mutants having changes in the C-terminal region of the proteolipid fail to show the presence of subunit 9 polypeptides in the immunoprecipitates, indicating the importance of the C-terminal region, extending from about residue 65, in the assembly of this protein. One other mutant that fails to assemble subunit 9 has a Gly_{18}|Asp substitution (strain 5102) in the N-terminal transmembrane region. On the other hand, strain 15B2 in which an aspartate is substituted at Gly_{23}, has no effect on the assembly of this protein. Both these glycine residues are highly conserved in proteolipids from a variety of organisms. This contrast in the assembly behaviour of these two mutants having the same substitution five residues apart, highlights the importance of residue 18 for the assembly of subunit 9, and suggests that its substitution by aspartate has serious consequences for the overall folding of the protein and perhaps its insertion into the membrane (see above).

The hydrophilic loop region of the proteolipid is suggested to play an important structural role in the function of the F_o-sector by making contacts with the F_1-sector (Sebald and Hoppe, 1981). Only a single *mit⁻* mutation in this region has been obtained; this mutant (strain 2422) carries a methionine substitution for an arginine residue which is conserved in proteolipids from all species. Loo et al., (1983) have implicated this arginine (Arg41 in *E. coli*, Arg39 in yeast) as being critical for the binding of the F_1-sector. In mutant 2422, however, the F_1 must be bound to F_o since the H^+-ATPase complex is fully assembled, suggesting this amino acid has a more significant role in the function of the complex rather than assembly.

The isolation of a series of intragenic revertants of mutant 2422 has allowed us to study the role of the hydrophilic loop region in more detail (Willson and Nagley, 1987). Data obtained with a same-site revertant (2422-R1) and two second-site revertants (2422-R9 and 2422-R10) is given in Table I. In strain 2422-R1, methionine is replaced by lysine, returning a positive charge to position 39. In strains 2422-R9 and

2422-R10, the methionine at position 39 is retained but a positive charge is introduced at position 35 by a Asn|Lys substitution. Strain 2422-R10 has an additional mutation (Leu_{53}|Phe), resulting in an oligomycin resistant growth phenotype. Biochemical analyses of these strains show that the ATPase activity has been restored to wild-type levels and that this activity is sensitive to oligomycin. The three revertants have high rates of respiration and mitochondrial respiratory control ratios are restored to near wild-type values.

The observations with *mit*⁻ strain 2422 and its revertants makes it clear that a positive charge, and not arginine *per se*, is essential in the region of residues 35 to 39. The location of a positive charge at residue 35, restoring activity, suggests this residue may be located in the loop region and not in the N-terminal transmembrane arm. Since we have not yet established whether the *mit*⁻ mutant 2422 is able to translocate protons through the F_o-sector it cannot be resolved whether the positive charge is essential for proton translocation or for coupling of this activity to ATP synthesis on the F_1-sector. A role for the loop region in coupling has been proposed (Mosher et al., 1985) based on a study in *E. coli* of mutant *unc*E114 that carries the amino acid substitution Gln_{42}|Glu in the loop region of subunit c (equivalent to Gln_{40} in yeast subunit 9).

MOLECULAR GENETICS OF SUBUNIT 6

Subunit 6 is the product of the *oli2* gene (Roberts et al., 1979) and is predicted from the gene sequence to be 259 amino acids in length (Macino and Tzagoloff, 1980; Novitski et al., 1984). Amino acid sequence comparisons of subunit 6 from diverse organisms indicate that the protein has two highly conserved regions at the C-terminal end, these being residues 155-197 (region I) and 231-254 (region II) in *S. cerevisiae* (see Novitski et al., 1984 for discussion). Subunit 6 is thought to span the inner mitochondrial membrane several times. Secondary structure predictions suggest five to seven transmembrane helices (Cox et al., 1986; Walker et al., 1984; Senior, 1983) and all predictions indicate that most of the two highly conserved regions span the mitochondrial membrane.

Molecular analyses of mutational changes in an extensive collection of *oli2 mit*⁻ mutants have recently been completed in this laboratory (John et al., 1986; John and Nagley, 1986). Two mutants (M11-28; C58, a temperature conditional *mit*⁻) were shown to have full-length *oli2* gene products with amino acid substitutions. All other *oli2 mit*⁻ mutants sequenced to date have truncated *oli2* gene products, most arising from frameshift mutations leading to premature termination. In the present study, we have analysed a selection of *oli2* mutants to identify domains required for function and assembly of the mtATPase complex.

The mitochondrial translation products present in the selection of *mit⁻* mutants used in this study are shown in Fig. 1. The electrophoretic mobilities in SDS-polyacrylamide gels of cytochrome oxidase subunit III and ATPase subunit 6 are very similar and their separation is not always possible. Subunit 6 is the lower of the two bands with electrophoretic mobility of 20 kDa, indicating that subunit 6 migrates faster than predicted from its amino acid sequence, which is 28 kDa. All *mit⁻* ᵐutants show the absence of the wild-type subunit 6 band in the display of their mitochondrial translation products. The *oli2* gene product of strain M11-28 has a slower mobility than that of the wild-type strain and co-migrates with cytochrome oxidase subunit III. It has been observed previously (Stephenson et al., 1981), using two-dimensional gel electrophoresis, that subunit 6 of strain M11-28 has an altered pI, probably a result of the introduction of the positive charge at amino acid 248 (John et al., 1986).

Strains Mb12, Ma1, M59-23 and M67-2 (tracks c,d,e and g) show the presence of shorter *oli2* translation products while no products are seen for strains M44 and M10-7 (tracks f and h). There is a reasonable correlation between the size of the truncated polypeptides predicted from gene sequence and their observed mobility relative to the anomalous mobility of the full-length subunit 6. The smaller the polypeptide the closer are the predicted and observed sizes (John et al., 1986). For example, strain M67-2 has a predicted subunit 6 derivative of 12.6 kDa and a protein with mobility of 11 kDa is observed. No predicted *oli2* gene product of 10 kDa or less has been observed on our gel system and thus no subunit 6 polypeptide is seen for strain M10-7 (predicted size 6.8 kDa; track h). The reason for the absence in strain M44 of an *oli2* gene product, predicted to be an 18.8 kDa protein with a mobility of approximately 13 kDa, is not clear. The novel C-terminus of this predicted protein is highly charged, containing four histidines and one glutamic acid. This high proportion of charged residues may prevent the protein from being inserted into the membrane, rendering it more susceptible to proteolytic digestion.

ASSEMBLY OF SUBUNIT 6 INTO H⁺-ATPase COMPLEX

The availability of a number of *mit⁻* mutants carrying truncated subunit 6 polypeptides of varying lengths provides an opportunity to determine the minimum length required for the assembly of subunit 6 with the H⁺-ATPase complex. The subunit 6 polypeptides associated with the H⁺-ATPase complex, as indicated by immunoprecipitation with the anti-β monoclonal antibody, are shown in Fig. 2 for

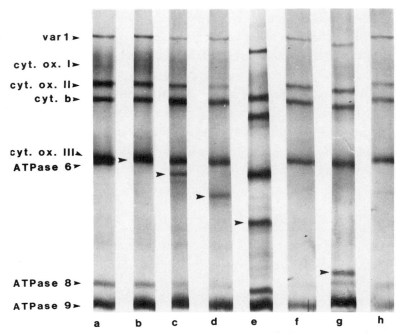

Fig. 1 Mitochondrial Translation Products of *oli2 mit⁻* Mutants. Cells were labeled *in vivo* with [³⁵S]sulphate in the presence of cycloheximide, and mitochondria were isolated (Choo et al., 1985). Mitochondrial proteins were solubilised in SDS, electrophoresed on 12% polyacrylamide gels in the presence of SDS and mitochondrial translation products were visualized by autoradiography. Autoradiogram illustrates mitochondrial translation products from strains (a) J69-1B, (b) M11-28, (c) Mb12, (d) Ma1, (e) M59-23, (f) M44, (g) M67-2, (h) M10-7. Subunit 6 polypeptides present in the mutant strains are indicated by arrows. The predicted length of these products and amino acid residues forming novel C-terminal sequences (first changed residue number in brackets) are as follows (John et al., 1986):

Mb12	246	(234)STLVLSNTMFDLS
Ma1	211	(210)FN
M59-23	178	(148)WLSITHIIRTCWYTITISTFISYYWNFILYC
M44	163	(143)VYMNMVEYSSHYSYTTVHHYH
M67-2	110	(102)LCLFLLTI
M10-7	57	(38)IIFIMYYYCIISYYKFMSIN

Strain M11-28 has a full-length subunit 6 product with two amino acid substitutions; Ser241|Phe and Thr248|Lys.

strains J69-1B, M11-28, Mb12, Ma1, M59-23, M67-2 and M10-7. Each of the mutant strains show the presence of novel polypeptides with mobilities corresponding to those observed in the mitochondrial translation products (Fig. 1). The other two F_o-subunits 8 and 9 are also assembled in each of the mutants. It is evident that subunit 6 polypeptides having at least 110 residues of the N-terminal region can be assembled into the H^+-ATPase complex.

The truncated subunits 6 present in the immunoprecipitates from strains M11-28 and Ma1 are more abundant than from strains Mb12, M59-23 and M67-2. These latter strains have extended novel regions at their C-termini, often with the introduction of charged residues in potential transmembrane segments (e.g. strains M59-23 and Mb12). The amount of the novel subunit 6 polypeptides present in immunoprecipitates can be correlated with the extent and nature of the amino acid substitutions.

The role of subunit 6 in the interaction of the F_o and F_1 sectors in the assembly of the H^+-ATPase complex can be addressed by consideration of the data obtained with strain M10-7 and M44. In these mutants the predicted subunit 6 polypeptide of 57 and 163 residues, respectively, are not observed in the gel display of mitochondrial translation products (tracks h and f, Fig. 1). Likewise, no novel peptides are displayed in immunoprecipitates of these strains (data for M44 not shown), and yet subunits 8 and 9 are associated with the F_1-sector precipitated with the anti-β antibody. The simplest conclusion is that subunit 6 is not required at all for the assembly of subunits 8 and 9 to the F_1-sector. It has been shown that in an *E. coli* mutant having the N-terminal 111 amino acids of subunit a (equivalent to yeast subunit 6) F_1 does bind to the F_o-sector, whereas no such binding was found for a second mutant having only the first 20 amino acid residues (M. Futai and M. Maeda, personal communication). This result suggests a role for subunit a, specifically between residues 20 and 111, in the interaction of the bacterial F_1 with F_o. It is possible that subtle differences exist between *E. coli* and yeast mitochondria in respect to the roles of subunit a and 6, respectively, in assembly of the F_1F_o complexes.

FUNCTIONAL ASPECTS OF SUBUNIT 6

What information concerning the function of subunit 6 in the H^+-ATPase can be obtained from our collection of mutants? All the *oli2 mit⁻* mutants lack ATP-P_i exchange activity and their ATP hydrolase activities are insensitive to oligomycin. This phenotype even applies to strain M11-28 having two amino acid substitutions in the second conserved transmembrane C-terminal region and strain Mb12 which lacks only 13 amino acids at the C-terminus. We conclude, therefore, that the second conserved

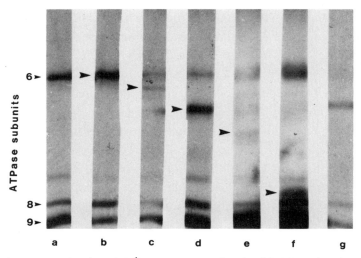

Fig. 2.   F$_o$-subunits of H$^+$-ATPase Associated with Monoclonal Anti-β Antibody Immunoprecipitates of *oli2 mit$^-$* Mutants.   Mitochondria isolated from cells labeled *in vivo* (Fig. 1) were solubilised with Triton X-100 and the H$^+$-ATPase complex was immunoprecipitated with monoclonal anti-β antibody RH48.6 coupled to Sepharose-4B beads as described previously (Hadikusumo et al., 1984).   Mitochondrial translation products associated with the antibody were visualized by fluorography after electrophoresis on a 12.5% SDS-polyacrylamide gel.   Tracks (a) J69-1B, (b) M11-28, (c) Mb12, (d) Ma1, (e) M59-23, (f) M67-2, (g) M10-7. Arrows indicate subunit 6 polypeptides.

transmembrane region of subunit 6, although not required for assembly, is critical for either proton translocation or for functional coupling of F_1 and F_o.

Full-length mutants with single amino acid substitutions provide the most useful information concerning functional domains within a protein. Only two such *oli2* mutants have been isolated to date in this laboratory; the *mit⁻* mutant M11-28 and the temperature-sensitive *mit⁻* mutant, C58. By analysis of revertants of strain M11-28, it was shown that the lesion resulting in loss of function in M11-28 is the introduction of the positive charge resulting from the substitution Thr248|Lys (John et al., 1986) in the second conserved transmembrane region. On the other hand, strain C58   the amino acid substitution Ser175|Tyr (John and Nagley, 1986) which lies in the first conserved transmembrane region. Strain M11-28 cannot grow on a non-fermentable substrate while strain C58 is able to utilise ethanol as an energy source at its permissive temperature, although to a lesser extent than the wild-type strain. Although this difference between M11-28 and C58 may only reflect the nature of the substitution, it would be of interest if one could assign a difference in functional significance to the two conserved regions.

In this respect, Cox et al. (1986) have recently suggested that the proton channel of *E. coli* (or mitochondrial) H⁺-ATPase is a hydrogen-bonding circuit in which subunit a (or 6) participates; for yeast this would involve amino acids 170-194 of the first conserved region of subunit 6. If the functional defect could be shown to be proton translocation, then the phenotype of M11-28 relative to C58 would imply that the second conserved transmembrane region is more important for proton channel activity. Indeed, Cain and Simoni (1986) have analysed a series of *E. coli* mutants with mutations in the two conserved regions and have shown that amino acid substitutions in the second conserved region totally abolished proton conductance while substitutions in the first conserved region had a less detrimental effect.

PERSPECTIVES

In this communication we have described how molecular characterization of mitochondrial gene mutations has been used to obtain information regarding structure-function relationships in two membrane-associated subunits of the F_o-sector of yeast H⁺-ATPase (subunits 6 and 9). The results also pointed to a role for subunit 9 in the development of other functional mitochondrial membrane complexes such as cytochrome oxidase.

In extending these studies attention will be given to the isolation and characterization of revertant mutations or mutations suppressing selected *mit⁻*

mutations. The usefulness of intragenic revertants in obtaining amino acid substitutions in particular domains of individual subunits has been discussed above. Intergenic suppressor mutations, on the other hand, could occur in either nuclear genes or other mitochondrial genes and would be expected to affect subunits which interact with the original defective F_o-subunit. For example, we have recently isolated a nuclear suppressor (Willson and Nagley, 1987) of mit^- strain 2422 (which carries the $Arg_{39}\backslash Met$ substitution in the hydrophilic loop of subunit 9). This nuclear mutant provides the opportunity to identify a putative nuclear-encoded subunit which interacts directly with the hydrophilic loop region of subunit 9, and could provide information on the important question of the binding of the F_1-sector with the F_o-sector. This approach is being extended to the isolation of nuclear suppressors of other mit^- mutants having mutations in either the *oli1*, *oli2* or *aap1* genes.

The detailed resolution of the function of any particular protein domain can best be achieved by a program of site-directed mutagenesis for specific amino acid substitutions. A strategy to achieve this goal in regard to mitochondrial gene products, in particular subunits 8 and 9, has recently been developed in this laboratory (Farrell et al., this volume). This strategy is based on the construction of synthetic genes designed for expression outside of the mitochondrion, where the cytoplasmically synthesized products have the necessary pre-sequences to deliver the subunit into the organelle leading to its functional assembly into the H^+-ATPase complex. This development offers the potential for more detailed analysis of structure-function relationships in subunits of the F_o-sector. Strategies are thus emerging for the controlled manipulation of the composition of the inner mitochondrial membrane, using normal or specifically mutated versions of F_o-sector subunits, thereby allowing new approaches to the study of the H^+-ATPase as well as other mitochondrial membrane complexes influenced by these subunits.

REFERENCES

Cain, B.D. and Simoni, R.D. (1986) J. Biol. Chem. 261, 10043-10050.
Choo, W.M., Hadikusumo, R.G. and Marzuki, S. (1985) Biochim. Biophys. Acta 806, 290-304.
Cox, G.B., Fimmel, A.L., Gibson, F., and Hatch, L. (1986) Biochim. Biophys. Acta 849, 62-69.
Hadikusumo, R.G., Hertzog, P.J., and Marzuki, S. (1984) Biochim. Biophys. Acta 765, 258-267.
Jean-Francois, M.J.B., Hadikusumo, R.G., Watkins, L.C., Lukins, H.B., Linnane, A.W. and Marzuki, S. (1986) Biochim. Biophys. Acta 852, 133-143.
John, U.P. and Nagley, P. (1986) FEBS Lett. 207, 79-83.

John, U.P., Willson, T.A., Linnane, A. and Nagley, P. (1986) Nucleic Acids Res. 14, 7437-7451.

Linnane, A.W., Lukins, H.B., Nagley, P., Marzuki, S., Hadikusumo, R.G., Jean-Francois, M.J.B., John, U.P., Ooi, B.G., Watkins, L., Willson, T.A., Wright, J. and Meltzer, S. (1985) in *Achievements and Perspectives of Mitochondrial Research*, Vol I, *Bioenergetics* (Quagliariello, E., Slater, E.C., Palmieri, F., Saccone, C., and Kroon, A.M., Eds) pp 211-222, Elsevier Science Publishers, Amsterdam.

Loo, T.W., Stan-Lotter, H., Mackenzie, D., Molday, R.S. and Bragg, P.D. (1983) Biochim. Biophys. Acta 733, 274-282.

Macino, G. and Tzagoloff, A. (1980) Cell 20, 507-517.

Macreadie, I.G., Choo, W.M., Novitski, C.T., Marzuki, S., Nagley, P., Linnane, A.W. and Lukins, H.B. (1982) Biochem. Int. 5., 129-136.

Macreadie, I.G., Novitski, C.E., Maxwell, R.J., John, U., Ooi, B.G., McMullen, G.L., Lukins, H.B., Linnane, A.W. and Nagley, P. (1983) Nucleic Acids Res. 11, 4435-4451.

Marzuki, S. and Linnane, A.W. (1979) Methods Enzymol. 56, 568-577.

Marzuki, S., Hadikusumo, R.G., Choo, W.M., Watkins, L., Lukins, H.B. ad Linnane, A.W. (1983) in *Mitochondria 1983: Nucleo-Mitochondrial Interactions* (Schweyen, R.J., Wolf, K. and Kaudewitz, F., Eds.) pp 535-549, Walter de Gruyter, Berlin.

Mosher, M.E., White, L.K., Hermolin, J., and Fillingame, R.H. (1985) J. Biol. Chem. 260, 4807-4814.

Nagley, P., Hall, R.M. and Ooi, B.G. (1986) FEBS Lett. 195, 1549-163.

Novitski, C.E., Macreadie, I.G., Maxwell, R.J., Lukins, H.B., Linnane, A.W., and Nagley, P. (1984) Curr. Genet. 8, 135-146.

Ooi, B.G., McMullen, G.L., Linnane, A.W., Nagley, P. and Novitski, C. (1985) Nucleic Acids Res. 13, 1327-1339.

Roberts, H., Choo, W.M., Murphy, M., Marzuki, S., Lukins, H.B. and Linnane, A.W. (1979) FEBS Lett. 108, 501-504.

Sebald, W., and Hoppe, S. (1981) Curr. Top. Bioenerg. 12, 1-64.

Senior, A.E. (1983) Biochim. Biophys. Acta 726, 81-95.

Stephenson, G., Marzuki, S., and Linnane, A.W. (1981) Biochim. Biophys. Acta 636, 104-112.

Walker, J.E., Saraste, M., and Gay, N.J. (1984) Biochim. Biophys. Acta 768, 164-200.

Willson, T.A., Ooi, B.G., Lukins, H.B. Linnane, A.W., and Nagley, P. (1986) Nucleic Acids Res. 14, 8228.

Willson, T.A., and Nagley, P. (1987) Eur. J. Biochem. in press.

THE STRUCTURE AND EXPRESSION OF A HUMAN GENE FOR

A NUCLEAR-CODED MITOCHONDRIAL ADENOSINE TRIPHOSPHATE

SYNTHASE BETA SUBUNIT

Shigeo Ohta, Hideaki Tomura, Kakuko Matsuda,
Kiyoshi Hasegawa* and Yasuo Kagawa

Department of Biochemistry, Jichi Medical School
Minamikawachi-machi, Tochigi-ken, 329-04, Japan

ABSTRACT

The beta subunit of mitochondrial ATP synthase is coded on a nuclear genome, synthesized in the cytosol, and then assembled with the other subunits which are coded on the mitochondrial genome. To determine the molecular mechanism by which the expressions on these two genetic systems are coordinated, the gene structure of the human ATP synthase beta subunit was determined, and the structure involved in this expression was analyzed. The gene for the beta subunit was found to be composed of 10 exons. The first exon corresponded to the prepiece peptide for targeting mitochondria. The 5' upstream region contained three CAT boxes (CCAAT), three GC boxes (CCGCCC) and four repeating sequences, but no typical TATA box. To determine the regulatory structure of the upstream region, fragments of various length were fused with a chloramphenicol acetyltransferase (CAT) gene and then transfected into a cultured cell. A 300 base pairs fragment was sufficient for expressing the CAT activity, and furthermore, the longer fragment (1300 base-pairs) enhanced the expression markedly. This result suggests that the gene for the human ATP synthase beta subunit has a regulatory structure. In addition, a restriction length fragment polymorphism in the gene and an interesting pseudo-gene are reported.

---

*Present address: Department of Pediatrics, Tohoku University School of Medicine, 1-1 Seiryo-cho, Sendai, 980 Japan

INTRODUCTION

Mitochondrial ATP synthase (F_0F_1) catalyzes ATP formation using the energy of the proton flux through the inner membrane in oxidative phosphorylation (Kagawa, 1984). The beta subunit is coded on a nuclear genome, synthesized in the cytosol and then assembled with other subunits which are coded on the mitochondrial genome (Hay et al., 1984). The number of mitochondria per cell varies considerably in different types of tissues and sometimes their numbers increase depending on the external conditions (Williams, 1986). Therefore, there should be some mechanisms which coordinate the gene expressions in these two genetic systems. As an approach to this problem, the gene structure for the beta subunit of the ATP synthase was analyzed. For these studies, we used a mammalian cultured cell line system since genes can be transfected into nucleus by using relatively simple methods (Gorman et al., 1982). Furthermore, the nucleo-mitochondrial interaction might be stricter in the mammalian system. As a target, we used a human gene because the mitochondrial DNA of this system is one of the smallest and has been well studied (Anderson et al., 1981).

In this report, we show the gene structure for the human ATP synthase beta subunit and the existence of a regulatory structure for gene expression. In addition, a restriction length fragment polymorphism (RLFP) and an interesting pseudo-gene structure are reported.

RESULTS AND DISCUSSIONS

cDNA Cloning for the Human ATP Synthase Beta Subunit (Ohta and Kagawa, 1986)

As the first step, cDNA for the human beta subunit was cloned from a lambda gt11 expression vector library (Young and Davis, 1983). The amino acid sequences of the beta subunits are highly conserved in various kinds of organisms. Therefore, the antibody could be used to detect the expressed protein for cloning the gene from the different organisms; the antibody against yeast beta subunit enabled us to clone the human gene. The deduced amino acid sequence of the human beta subunit was 97% conserved with the bovine one which had been determined by protein sequencing (Renswick and Walker, 1983). The prepiece peptide, which is shown in Fig. 1, is rich in arginine and serine as those of the other mitochondrial proteins are (Hay et al., 1984).

Cloning of the Genomic Gene for the Human ATP Synthase Beta Subunit and the Gene Structure

The genomic gene was cloned by screening a human genomic Charon 4A library using a cloned cDNA as the hybridization probe. The exon/intron junctions were determined by direct comparison with the nucleotide sequence for the cDNA. Fig. 2 shows a schematic drawing of the gene. The gene was composed of 10 exons and the first exon corresponds to the prepiece peptide (Fig. 1) although the prepiece did not construct its own exon in a tobacco gene (Boutry and Chua, 1985). The junctions of the exons and the introns were quite different from those of the tobacco gene (Boutry and Chua, 1985).

The 4th and 5th exons contain some regions that are homologous to some nucleotide binding proteins such as ras gene product (Yuasa et al., 1983), myosine (Walker et al., 1982) or ADP/ATP translocator (Aquila et al., 1982). These exons were considered to compose a functional domain.

Four long repeating sequences were found in the 5' upstream region and the 3rd and the 9th introns and an inverted repeat sequence was found in the 8th intron. The 5' upstream region contained three CAT boxes (CCAAT) (Benoist et al., 1980) and three GC boxes (CCGCCC) (Kadonaga et al., 1986), but no typical TATA box (Brathnach and Chambon, 1981).

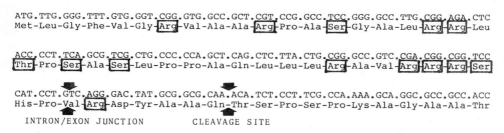

Fig. 1. Nucleotide Sequence and the Deduced Amino Acid Sequence for the Prepiece Peptide for Targeting to Mitochondria. The basic and the hydroxyl amino acids are boxed. The junction point of the intron/exon and the cleavage site are shown by arrows.

Restriction Fragment Length Polymorphism (RFLP)

A restriction fragment length polymorphism was found by the total Southern blotting of 27 unrelated Japanese individuals and 12 members in three generations of one family. RFLP characterized by polymorphic bands at 3.2, 1.7 and 1.5 kilobase-pairs was detected in *Pst*I digests of the DNAs. This finding will allow us to employ this gene as a human genetic marker using RFLP.

A Pseudo-gene Lacking 4 Exons

In the total Southern blotting with various kinds of restriction enzymes, some fragments that hybridized with the cDNA did not fit the restriction map for the bonafide gene. Actually another gene that hybridized with the cDNA was cloned from the same genomic library. The nucleotide sequence showed that this second gene had the introns but this gene lacked the 4th to the 7th exons. Since the 5th and the 6th exons corresponded to the nucleotide binding site, these missing exons might have been transferred to other nucleotide binding genes such as myosine, ras gene product or ADP/ATP translocator during evolution.

The Regulatory Structure of the Human Beta Gene

To examine the promoter activity, the 5' upstream region was fused to a chloramphenicol acetyltransferase (CAT) gene (Laimins et al., 1982) and was transfected to Hela cells by the calcium phosphate precipitation method (Davis et al., 1986). To ligate fragments of various lengths to the CAT gene, the DNA fragment for the upstream region was digested by *Bal*31 exonuclease. The transfected cells were harvested after 48 hours and lysed by sonication. The CAT activity expressed in the cells was measured by separation of acetylated 14-C chloramphenicol on a thin layer chromatography (Gorman et al., 1982). Fig. 3 shows the expression of the CAT activity. When 300 base-pairs were fused to the CAT gene, the activity was detected at a level comparable with that obtained by fusion of 600 base-pairs (bp). Furthermore, the 1300 bp fragment enhanced the expression markedly as shown in Fig. 3. The upstream region of 300 bp contained three CAT boxes (CCAAT), and the upstream region of 1300 bp contained two GC boxes (CCGCCC) and three repeating sequences.

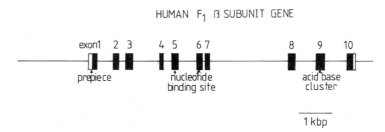

HUMAN F$_1$ β SUBUNIT GENE

Fig. 2. A Schematic Drawing of the Full–Length of the Human ATP Synthase Beta Subunit Gene. The closed and the open boxes show the coding region and the uncoding region, respectively. The line shows the introns.

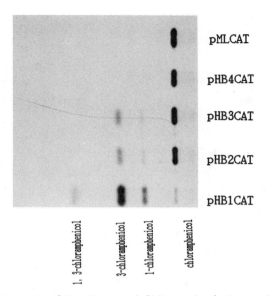

pMLCAT

pHB4CAT

pHB3CAT

pHB2CAT

pHB1CAT

1, 3-chloramphenicol

3-chloramphenicol

1-chloramphenicol

chloramphenicol

Fig. 3. Measurement of the Expressed Chloramphenicol Acetyl Transferase Activity. The 5' upstream region fragment was digested with *Bal*31 exonuclease and ligated to a plasmid harboring the CAT gene. The plasmids were transfected into Hela cells by the calcium phosphate coprecipitation method. pHB1CAT has the 1300 bp upstream region including 3 GC boxes, three repeating sequences and three CAT boxes. pHB2CAT has the 600 bp upstream region including a GC box and three CAT boxes. pHB3CAT has 300 bp including three CAT boxes. pHB4CAT has 30 bp. The pMLCAT without any upstream region was used as a negative control. The HeLa cells were harvested and lysed after 48 hours, and then the lysates were incubated with [^{14}C]chloramphenicol and 1 mM acetyl CoA for 1 hour. The resulting acetylated chloramphenicol was separated by thin layer chromatography and visualized by autoradiography.

These results suggest the existence of an enhancing element(s) in the upstream region between 600-1300 bp. In addition, when the same plasmids were transfected into a mouse L-cell line, good expression of the CAT activity was observed (data not shown). This indicates that the promoter region are common to the mouse gene. Here, we showed the gene structure and the existence of the regulatory structure for the expression. Further studies are necessary to determine the fine structure for the regulation, and in principle, it is possible to isolate the regulatory factor.

REFERENCES

Anderson, S., Bankier, A.T., Barrell,B.G., de Bruijin, M.H.L., Coulson, A.R., Drouin, J., Eperon, I.C., Nierlich, D.P., Roe, B.A., Sanger, F., Schreier, P.H., Smith, A.J.H., Staden, R. and Young, I.G. (1981) Nature 290, 457-465.

Aquila, H., Misra, D., Eulitz, M. and Klingenberg, M. (1982) Hoppe-Seyler's Z. Physiol. Chem. 363, 345-349.

Benoist, C., O'Hara, K., Breathnach, R. and Chambon, P. (1980) Nucleic Acids Res. 8, 127-142.

Boutry, M. and Chua, N-H. (1985) EMBO J. 4, 2159-2166.

Brathnach, R. and Chambon, P. (1981) Ann. Res. Biochem. 50, 349-383.

Davis, L.G., Dibner, M.D. and Batter, J.F. (1986) Basic Methods in Molecular Biology, Elsevier, Amsterdam.

Gorman, C., Moffat, L. and Haward, B. (1982) Mol. Cell. Biol. 2, 1044-1051.

Hay, R., Boehni, P. and Gasser, S. (1984) Biochim. Biophys. Acta. 779, 65-87.

Kagawa, Y. (1984) in Bioenergetics (Ernster, L., ed.) pp. 149-186, Elsevier, Amsterdam.

Kadonaga, J.T., Jones, K.A. and Tjian, R. (1986) Trends Biochem. Sci. 11, 20-23.

Laimins, L., Khoury, G., Gorman, C., Howard, B.H. and Gruss, P. (1982) Proc. Natl. Acad. Sci. USA 79, 6453-6457.

Ohta, S. and Kagawa, Y. (1986) J. Biochem. (Tokyo) 99, 135-141.

Renswick, M.J. and Walker, J.E. (1983) J. Biol. Chem. 258, 3081-3089.

Walker, J.E., Saraste, M., Runswick, M.J. and Gay, N.J. (1982) EMBO J. 1, 945-951.

Williams, R.S. (1986) J. Biol. Chem. 261, 12390-12394.

Young, R.A. and Davis, R.W. (1983) Proc. Natl. Acad. Sci. USA 80, 1194-1198.

Yuasa, Y., Srivastava, S.K., Dunn, C.Y., Rhim, J.S., Reddy, E.P. and Aaronson, S.A. (1983) Nature 303, 775-779.

ISOLATION OF A cDNA CLONE FOR THE γ SUBUNIT OF

THE *CHLAMYDOMONAS REINHARDTII* CF$_1$

Lloyd M. Yu, Steven M. Theg and Bruce R. Selman

Department of Biochemistry, College of Agricultural and Life Sciences
University of Wisconsin-Madison, Madison, WI 53706

ABSTRACT

A cDNA library from *Chlamydomonas reinhardtii* was probed with antiserum directed against the nuclear encoded γ subunit of the chloroplast H$^+$-translocating ATP synthase (CF$_0$-CF$_1$) of *C. reinhardtii*. A cDNA was isolated and transcribed *in vitro*. The transcript was translated *in vitro* and immunoprecipitated with anti-γ serum to yield a product that co-electrophoresed with the immunoprecipitated product from *in vitro* translated polyadenylated RNA. These proteins were larger than the mature γ subunit. Thus the γ subunit is synthesized as a precursor of greater molecular weight in *C. reinhardtii*, as has been suggested for the γ subunit from spinach (Nelson et al., 1980).

The precursor protein encoded by this cDNA was imported into pea chloroplasts and processed to a lower molecular weight polypeptide that co-electrophoresed with mature *C. reinhardtii* γ subunit. The largest cDNA isolated is about the same length as the corresponding mRNA (about 2000 bases). Southern analyses revealed restriction fragment length polymorphisms and that the γ subunit is probably encoded by an intron-containing single copy gene.

INTRODUCTION

Proton-translocating ATP synthases (ATPases) are a class of multi-subunit enzymes found in energy transducing membranes. The CF$_0$-CF$_1$ is the thylakoid-bound protein complex that catalyzes ATP synthesis at the expense of the pH gradient generated during photosynthetic electron transfer. The complex contains nine different subunits (Hennig and Herrmann, 1986; Vignais and Satre, 1984) of which three (subunits γ and δ of the CF_1, and subunit II of CF_o) are synthesized in the cytosol and probably encoded by nuclear DNA (Nelson, 1981). The initial translation products for these subunits presumably bear amino terminal transit sequences (Chua and Schmidt, 1979; Schmidt and Mishkind, 1986) which are cleaved during import into the chloroplast. The ability to synthesize the nuclear-encoded subunits of the CF_o-CF_1 *in vitro* should aid in the study of the assembly of the holoenzyme, the possible turnover of the individual subunits in the mature enzyme (Merchant and Selman, 1984), and also, the general mechanism of protein import into the chloroplast. These studies are most easily accomplished using isolated chloroplasts, and, thus, it becomes essential to synthesize chloroplast destined proteins *in vitro*. Toward this end we have constructed a cDNA library from *C. reinhardtii* in the chimeric-protein expression vector λgt11 and cloned a cDNA for the precursor to the γ subunit of the CF_o-CF_1. While this work was in progress, Tittgen et al. (1986) published a similar study describing the isolation of a cDNA for the γ subunit of the spinach ATP synthase.

MATERIALS AND METHODS

Library Construction

A cDNA library of *C. reinhardtii* strain 2137 polyadenylated RNA was constructed in the phage expression vector λgt11 as previously described (Merchant and Bogorad, 1987).

Screening

The λgt11 library was screened with antibodies directed against the γ subunit of *C. reinhardtii* CF_1 after removal of *E. coli* cross-reactive antibodies (Young and Davis, 1985).

Preparation of Antigens and Antisera

C. reinhardtii CF_1 was purified as previously described (Selman-Reimer et al., 1981) with minor modifications. The protein was stored as an ammonium sulfate precipitate prior to use as an antigen. The γ subunits of CF_1 were separated by preparative gel electrophoresis and was electroeluted (Allington et al., 1978) from the appropriately excised portion of the gel after visualizing with KCl (Nelles and Bamburg, 1976). Antibodies to purified CF_1 and the γ subunit of CF_1 were raised in rabbits.

Plasmids

The cDNA inserts were prepared by *Eco*RI digestion of λgt11 DNA, agarose gel electrophoresis (Maniatis et al., 1982), and freeze/thaw rupture and centrifugation of excised gel fragments (Tautz and Renz, 1983). The inserts were ligated into the *Eco*RI site of the polylinker in the plasmid PTZ18R (United States Biochemical Corporation, Cleveland, OH) before transformation of the host cell JM101.

RNA Preparation

Total RNA and polyadenylated RNA were isolated from *C. reinhardtii* strain CC 124 as previously described (Merchant and Bogorad, 1986).

Hybridizations

After electrophoresis, nucleic acids were transferred to nylon filters by capillary action and fixed by UV irradiations (Church and Gilbert, 1984). Nick-translated probes (Bethesda Research Laboratories Inc., Gaithersburg, MD) containing [α-^{32}P]deoxycytidine were hybridized to the filters at 65-68°C (Church and Gilbert, 1984) and were visualized by autoradiography after washing.

Transcription/Translation

Plasmid templates were linearized with *Hind*III (Maniatis et al., 1982), and the run off *in vitro* transcription reaction was catalyzed by T7 RNA polymerase (Promega Biotec, Madison, WI). *In vitro* translation reactions were carried out in rabbit reticulocyte lysates (Bethesda Research Laboratories Inc., Gaithersburg, MD) at 88.5 mM K$^+$, 1.17 mM Mg^{2+}, and 20 μCi [^{35}S]methionine in 30 μl reaction volumes.

Immunoprecipitations

The immunoprecipitation of [^{35}S]Na$_2$SO$_4$ labeled cell extracts and *in vitro* translation reactions were performed essentially as described (Merchant and Bogorad, 1986). Crude *C. reinhardtii* CF$_1$ was prepared from ^{35}S-labeled thylakoid membranes by chloroform extraction; the high speed aqueous supernatant, brought to 200 mM NaCl, was immunoprecipitated with anti-CF$_1$ serum without the addition of detergents. The pellets of IgGsorb (The Enzyme Center, Boston, MA), which contained the antigen-antibody complexes, were washed with 50 mM Tris-Cl (pH 8.0), 200 mM NaCl, 1 mM EDTA, and 0.5% Nonidet P-40. Prior to gel electrophoresis, the pellets were resuspended and heated in sample buffer.

Cells: Growth, Labeling and Sample Preparation

C. reinhardtii strain CC 124 was grown photoheterotrophically to late log or early stationary phase on Tris-acetate-phosphate medium (Gorman and Levine, 1965) containing 200 μM SO_4^{2-} and 3 mCi $[^{35}S]Na_2SO_4$ in 125 ml of culture. The cells were pelleted at 4°C at 3000 x g for 3 min and washed once in 10 mM NaPi buffer (pH 7.1). A 30 μl sample, containing 50 μg of chlorophyll, was frozen at -80°C and thawed at 25°C twice before detergent solubilization and immunoprecipitation (Merchant and Bogorad, 1986). The remainder of the cells were used to prepare a CF_1 containing extract essentially as described above.

In vitro Import Into Pea Chloroplast

The preparation of pea chloroplasts and the *in vitro* import assays were performed as previously described (Cline et al., 1985).

RESULTS

cDNA Isolation

The first screening of the λgt11 library, containing 10^6 phage, gave twenty-five positive signals. These regions were excised from the agar plate and after three additional screenings fourteen purified phage remained. Of these, nine were arbitrarily chosen. A reverse Western (Young and Davis, 1985) demonstrated that these cDNAs, when expressed, were able to specifically remove the anti-γ antibodies from the anti-CF_1 serum (data not shown). DNA was isolated from these phage (Maniatis et al., 1982) and digested with *Eco*RI prior to electrophoresis. The three largest inserts (approximately 1900, 1760, and 1360 base pairs) were excised from the agarose gel and were ligated into the *Eco*RI site in the plasmid PTZl8R (and designated PTZ-γCF_1-1A, PTZ-γCF_1-4A, and PTZ-γCF_1-9A, respectively). These three plasmids were used to transform *E. coli* strain JM101.

Northern Analysis

The plasmids PTZ-γCF_1-1A, PTZ-γCF_1-4A, and PTZ-γCF_1-9A and the host plasmid (PTZ18R) were nick translated and used to probe total RNA from *C. reinhardtii* for hybridizing sequences (Fig. 1). The plasmids that contain cDNAs primarily hybridize to a single RNA species (lane 1 vs. lanes 2, 3, and 4). This RNA is approximately 1900 bases in length and is about the same size as the largest cDNA.

Genomic Southern Analysis

Nick translated cDNA (PTZ-γCF$_1$-lA) was used to probe restricted and electrophoresed genomic DNA from two strains of *C. reinhardtii*. Strain 2137 (Fig. 2, left) was the source of the cDNA library and was derived from a cross between strain CC 124 (Fig. 2, right) and strain 21 gr. The *Hind*III, *Pvu*II, and *Sst*I/*Sac*I banding patterns generated with strain 2137 DNA are consistent with the γ subunit being encoded by either a single-copy gene containing an intron or two closely linked gene copies. However, the *Hinf*I and *Pst*I derived banding patterns for strain 2137 are only consistent with the former possibility since too few hybridizing fragments are generated for closely-linked gene copies. A similar analysis of the banding patterns from the CC 124 digestions yields the same conclusion, namely the γ subunit is probably encoded by a single-copy gene that contains an intron.

Interestingly the Southern analyses of the DNA from strains 2137 and CC 124 reveal restriction fragment length polymorphisms. These patterns can be rationalized by two differences between the strains. An insertion, flanking the gene, present in

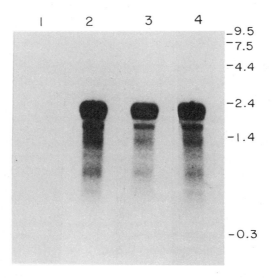

Fig. 1. Hybridization of Nick–Translated Plasmids to Total RNA Samples from *C. reinhardtii*. Total RNA samples were denatured and electrophoresed in an agarose gel containing formaldehyde which was then blotted and fixed to a nylon filter. The filter was divided and incubated with ^{32}P-labeled plasmid DNA produced by nick translation. Lane 1: host plasmid DNA (PTZ18R). Lane 2: plasmid PTZ-γCF$_1$-lA. Lane 3: plasmid PTZ-γCF$_1$-4A. Lane 4: plasmid PTZ-γCF$_1$-9A. The arrows on the right denote the approximate number of kilobases.

strain 2137 (and absent in strain CC 124) suffices to reconcile the differences in the lengths of hybridizing fragments generated by an endonuclease. The other difference can be ascribed to a second *Pst*I restriction site within the γ gene in strain 2137.

In vitro **Translation,** *in vitro* **Labeling and Immunoprecipitation**

Three linearized plasmids were transcribed *in vitro* and the transcription products were isolated and then translated *in vitro* in the presence of [^{35}S]methionine. The samples were electrophoresed in a 15% polyacrylamide gel before and after immunoprecipitation and ^{35}S-labeling was visualized by fluorography (Fig. 3). The vector (PTZ18R) yielded translation products identical to the minus RNA control (lanes 7 and 5); in neither case did treatment with anti-γ serum yield an immunoprecipitable product (lanes 15 and 13). The plasmids PTZ-γCF$_1$-1A and PTZ-γCF$_1$-1C contained the same cDNA insert but in opposite orientations. The translation product from PTZ-γCF$_1$-1C (lane 6) was not immunoprecipitated by anti-γ serum (lane 14). However, in the other orientation, the transcribed cDNA was translated to give essentially one product (lane 2) which was readily

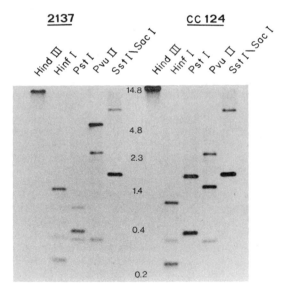

Fig. 2. Autoradiograph of Southern Hybridizations of Genomic DNA from *C. reinhardtii* Strains 2137 (left) and CC 124 (right) with a cDNA Probe of the γ Subunit. Genomic DNA (1 μg) was restricted overnight with various endonucleases, separated by gel electrophoresis, and then transferred to a nylon filter. The blot was probed with nick-translated plasmid PTZ-γCF$_1$-1A. This cDNA probe was derived from strain 2137 polyadenylated RNA. Within this cDNA, *Hind*III and *Sst*I/*Sac*I have no recognition sites, *Hinf*I and *Pst*I each have two recognition sites, and *Pvu*II has one recognition site. Size markers in kilobases are indicated in the center lane.

immunoprecipitated with anti-γ serum (lane 10). The major translation product derived from PTZ-γCF₁-1A comigrated with the product, immunoprecipitated with anti-γ serum, from *in vitro* translated polyadenylated RNA from *C. reinhardtii* (lanes 1 and 9). Both of these bands migrated above the mature γ subunit immunoprecipitated either from detergent extracted cells with anti-γ serum (lanes 3 and 11) or from a crude CF₁ containing extract with anti-CF₁ serum (lanes 4, 8, and 12).

Import Into Pea Chloroplasts

Reticulocyte lysates, containing ^{35}S-labeled pre-γ subunit, were incubated with pea chloroplasts *in vitro* and, after 20 min at 25°C, the mixtures were treated with

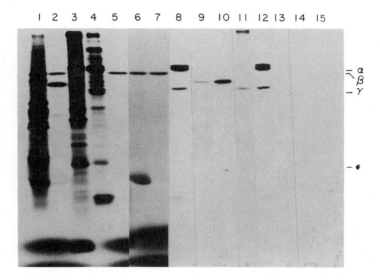

Fig. 3. Polyacrylamide Gel Electrophoretic and Fluorographic Analysis of *In Vitro* Translations of Transcribed cDNAs and Polyadenylated RNA and of *In Vivo* Labeled Cells. Lanes 1-7: total translates or cell extracts prior to immunoprecipitation. Lanes 8-15: immunoprecipitates of total translates or cell extracts. Lanes 1 and 8, 2 and 9, and 6 and 13: [^{35}S]methionine labeled *in vitro* translation reactions using transcription products from linearized plasmids PTZ-γCF₁-1C (lane 1), PTZ18R (lane 2), and PTZ-γCF₁-1A (lane 6) and the ^{35}S-labeled products immunoprecipitated from these *in vitro* translations with anti-γ serum (lanes 8, 9, and 13, respectively). Lanes 3 and 10, and 7 and 14: labeled products from *in vitro* translation containing [^{35}S]methionine with either no RNA (lane 3) or polyadenylated RNA (lane 7) and the ^{35}S-labeled products immunoprecipitated from these *in vitro* translations with anti-γ serum (lanes 10 and 14, respectively). Lanes 4, 11, and 15: a crude chloroform extract of *C. reinhardtii* membranes (lane 4) derived from cells labeled *in vivo* with [^{35}S]Na₂SO₄ and the ^{35}S-labeled products immunoprecipitated with anti-CF₁ serum (lanes 11 and 15). Lanes 5 and 12: ^{35}S-labeled products from whole cells, labeled *in vivo* with [^{35}S]Na₂SO₄, after detergent solubilization (lane 5) and immunoprecipitation with anti-γ serum (lane 12). The subunits of CF₁ are labeled on the far right.

protease or buffer. The chloroplasts were then re-isolated from a Percoll gradient, dissolved in sample buffer, and electrophoresed in a polyacrylamide gel. These results are summarized in Fig. 4 [Note that the reticulocyte lysate labeled, in addition to the pre-γ subunit (lane 3), an indigenous protein that migrates just below the β subunit of the *C. reinhardtii* CF_1 (lane 2)]. The pre-γ subunit associated with chloroplast, in a protease sensitive manner, regardless of the energy state of the import reaction (lanes 4, 6, and 8 vs lanes 5, 7, and 9). Compared to "energized" chloroplasts (chloroplasts either illuminated or supplemented with ATP) more protease sensitive precursor bound to dark incubated chloroplasts presumably because the subsequent import step

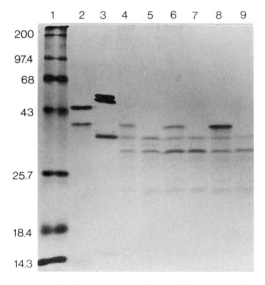

Fig. 4. Polyacrylamide Gel Electrophoretic and Fluorographic Analysis of the Import of the Pre-γ Subunit from *C. reinhardtii* into Pea Chloroplasts. PTZ–γCF_1-1A was transcribed *in vitro* and the transcription products were translated *in vitro* in a rabbit reticulocyte lysate in the presence of [^{35}S]methionine. The lysate was incubated with intact pea chloroplasts in the absence (lanes 6-9) or presence (lanes 4 and 5) of light. Half of the dark incubations were supplemented to 3 mM MgATP (lanes 6 and 7). After the incubation period, each sample was divided in half, and one portion was treated with thermolysin (lanes 5, 7, and 9) while the other portion was treated with buffer (lanes 4, 6, and 8). Intact chloroplasts were then re-isolated on Percoll gradients, solubilized in sample buffer and electrophoresed. Labeled products were revealed by fluorography. Lane 2 contains an immunoprecipitated (^{35}S-labeled) *C. reinhardtii* CF_1 standard, and lane 3 contains a sample of the reticulocyte lysate prior to incubation with the pea chloroplasts. The subunits of CF_1 are labeled on the far right. The kilodalton values for the molecular weight standards (lane 1) are labeled on the far left.

required more (Mg)ATP than was present (lane 8 vs lanes 4 and 6). Nevertheless, in the dark, a low level of import occurred (lane 9 vs lanes 5 and 7) since pre-existing pools of ATP were not intentionally depleted from either the chloroplasts or the translation reaction mixtures. However, in the presence of light, or additional (Mg)ATP, the precursor was cleaved to faster migrating species that were, unlike the bound precursor, resistant to exogenous protease treatment (lanes 4 and 6 vs lanes 5 and 7). The largest of the imported polypeptides comigrated with the mature γ subunit found in the *C. reinhardtii* CF_1 (lane 2). Whether an amino terminal transit sequence was actually cleaved from the precursor to yield the polypeptide that comigrated with the mature γ subunit has not yet been established.

DISCUSSION

We have isolated a cDNA for the pre-γ subunit of the thylakoid ATP synthase from *C. reinhardtii*. Our largest cDNA (PTZ-γCF_1-1A) probably contains the entire coding region for this nuclear encoded subunit and certainly contains sufficient information for the *in vitro* synthesized polypeptide to be transported into and processed by pea chloroplasts. The hybridization of nick translated cDNA to a sample of total RNA from *C. reinhardtii* shows that our largest cDNA is approximately the same size as the most strongly hybridizing RNA (Fig. 1).

Southern blot analysis of strain CC 124 and its F1 progeny strain 2137, from which the cDNA library was constructed, suggests that the pre-γ subunit is encoded by an intron containing single-copy gene (Fig. 2). Furthermore, the Southern analysis of these two strains reveals restriction fragment length polymorphisms consistent with a flanking insertion and the acquisition of a restriction site in strain 2137.

The electrophoretic mobilities of immunoprecipitated polypeptides translated *in vitro* or *in vivo* show that the γ subunit is synthesized as a larger precursor (Fig. 3). Our gel system cannot resolve proteins of only slightly differing molecular weights, thus it remains a possibility that our cDNA does not contain the entire coding region for the pre-γ subunit. Confirmation of the length and information content of our cDNA must await its sequence determination.

The precursor to the *C. reinhardtii* γ subunit is imported by pea chloroplasts (Fig. 4). This import is associated with the processing of the precursor to a product that comigrates with the mature γ subunit from *C. reinhardtii*. It appears, although we are not certain, that the amino terminal transit peptide is faithfully removed by the pea processing enzyme. Within the chloroplast, the locations of the imported, mature sized polypeptide and the smaller degradation products are unknown. Although the exact similarities between the γ subunits from pea and *Chlamydomonas* are unknown, they are to some degree homologous since rabbit antisera raised against the *C. reinhardtii* γ subunit cross reacts with the γ subunit from pea CF_1 (Western blot not shown).

ACKNOWLEDGMENTS

This work was supported in part by grants from the College of Agricultural and Life Sciences, University of Wisconsin-Madison and the National Institutes of Health (GM 31384). The authors also wish to thank Dr. Sabeena Merchant for the construction of the λgtll library.

REFERENCES

Allington, W.B., Cordry, A.L., McCullough. G.A., Mitchell, P.E. and Nelson, J.W. (1978) Anal. Biochem. 85, 188-196.

Chua, N.-.H. and Schmidt, G.W. (1979) J. Cell Biol. 81, 461-483.

Church, G.M. and Gilbert, W. (1984) Proc. Natl. Acad. Sci. USA 81, 1991-1995.

Cline, K., Werner-Washburne, M., Lubben, T.H. and Keegstra, K. (1985) J. Biol. Chem. 260, 3691-3693.

Gorman, D.S. and Levine, R.P. (1965) Proc. Natl. Acad. Sci. USA 54, 1665-1669.

Hennig, J. and Herrmann, R.G. (1986) Mol. Gen. Genet. 203, 117-128.

Maniatis, T., Fritsch, E.F. and Sambrook, J. (1982) In: Molecular Cloning: A Laboratory Manual, pp. 55-186, Cold Spring Harbor.

Merchant, S. and Bogorad, L. (1986) Mol. Cell. Biol. 6, 462-469.

Merchant, S. and Bogorad, L. (1987) J. Biol. Chem., in press.

Merchant, S. and Selman, B.R. (1984) Plant Physiol. 75, 781-787.

Nelles, L.P. and Bamburg, J.R. (1976) Anal. Biochem. 73, 522-531.

Nelson, N. (1981) Curr. Topics Bioenerg. 11, 1-33.

Nelson, N., Nelson, H. and Schatz. G. (1980) Proc. Natl. Acad. Sci. USA 77, 1361-1364.

Schmidt, G.W. and Mishkind, M.L. (1986) Ann. Rev. Biochem. 55, 879-912.

Selman-Reimer, S., Merchant, S. and Selman, B.R. (1981) Biochemistry 20, 5476-5482.

Tautz, D. and Renz, M. (1983) Anal. Biochem. 132, 14-19.

Tittgen, J., Hermans, J., Steppuhn, J., Jansen, R., Jansson, C., Andersson, B., Nechustai, R., Nelson, N. and Hermann, R.G. (1986) Mol. Gen. Genet. 204, 258-265.

Vignais, P.V. and Satre, M. (1984) Mol. Cell. Bioch. 60, 33-70.

Young, R.A. and Davis, R.W. (1985) In: Genetic Engineering: Principles and Methods (J.K. Setlow and A. Hollander, eds.) pp. 29-41, Plenum Press, New York and London.

IMPORT AND ASSEMBLY OF THE MITOCHONDRIAL ATP SYNTHASE OF BAKER'S YEAST

David Burns, Laura Hefta, David Modrak and Alfred Lewin

Indiana University, Department of Chemistry
Bloomington, Indiana 47405 USA

The synthesis of mitochondrial ATPase is programmed by nuclear genes as well as by mitochondrial genes. In *Saccharomyces cerevisiae*, the coding dichotomy parallels the structural dichotomy of the enzyme: The five F_1 subunits and OSCP are encoded in the nucleus, and the three known F_0 subunits, 6, 9, 10 (also called *aap*1), are encoded in the mitochondrion. Because formation of both domains is essential if new ATPase complex is to be made, we are investigating the assembly of ATPase and particularly the interaction between nuclear and mitochondrial gene products. Our experiments are of two types: (1) studies of the rate of import and association of the ATPase subunits which are encoded in the nucleus and made in the cytoplasm; and (2) examination of nuclear mutations (suppressors) which compensate for a defect in the mitochondrial gene for an F_0 subunit.

KINETICS OF ATPase SUBUNIT IMPORT

When precursors of mitochondrial proteins are made in a cell-free protein synthesis system and incubated with freshly isolated mitochondria, they are transported into the organelles and converted to their mature forms by proteolytic cleavage of their targeting peptides (Neupert and Schatz, 1981; Schatz and Butow, 1983). We have developed an assay for assembly of F_1-ATPase subunits using antibody co-precipitation as a criterion for F_1 formation from newly imported subunits (Lewin and Norman, 1983; Burns and Lewin, 1986). Following the import reaction, mitochondria are dissociated in non-ionic detergent, and antibodies to the α subunit are used to immunoprecipitate F_1. Co-precipitation of β and γ subunits which had been labeled during cell-free translation suggests that assembly of F_1 has occurred. We can also measure this association following import into mitochondria of strains which lack assembled F_1. We assume, therefore, that we are detecting assembly of newly-made subunits and does not indicate exchange of imported subunits with pre-existing enzyme.

In our current experiments, we compared the rates of import of individual F_1 subunits with the rates of their appearance in the assembled enzyme. The rate of import for the α and β subunits was determined by incubating labeled precursors made *in vitro* with mitochondria and then stopping the reaction after increasing intervals. The amount of import was determined by immunoprecipitation with subunit-specific serum. Samples were separated on polyacrylamide gels and quantitated by eluting the bands corresponding to α and β subunits and determining their radioactivity in a scintillation counter (Fig. 1). We discovered that the precursor to α subunit was imported and processed to its mature form much more rapidly than is the β precursor. From the data presented in Fig. 1, it appeared that α import was maximal by 6 minutes of incubation, but that β import did not reach a plateau until 60 minutes.

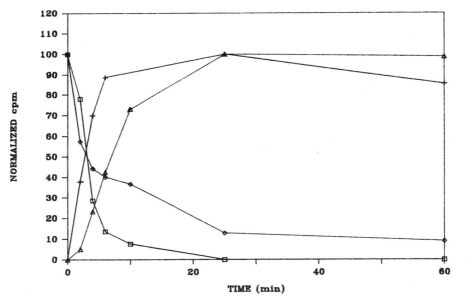

Fig. 1. α Subunit of F_1 ATPase is Imported More Rapidly than β Subunit. (+) α subunit in the mitochondrial pellet. (□) α precursor in the supernatant. (△) β subunit in the mitochondrial pellet. (◇) β precursor in the supernatant.

IMPORT OF INDIVIDUAL SUBUNITS

If the α subunit is indeed imported faster than the β subunit, we might expect to find a pool of unassembled ATPase subunits within the organelle *in vivo*. Kinetic labeling experiments (reported in Burns and Lewin, 1986) indicated no large pool of unassembled β or γ subunits *in vivo*. Nevertheless, these double-label experiments suffered from lack of sensitivity, and we have sought an alternative method to probe for small, transient pools of subunits. Our approach has been to produce individual ATPase subunits *in vitro* and to import these radiolabeled proteins into mitochondria independently of each other. Co-precipitation with antibody specific for the α subunit would suggest that unassembled α subunit was already present in the isolated mitochondria.

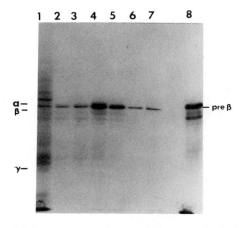

Fig. 2. Assembly of β Precursor Implies the Existence of a Pool of α Subunits. Lane 1: Import of F_1 subunits from translation of total yeast RNA, dissociation with Triton X-100, immunoprecipitated with α-specific antiserum; lane 2: import of pre-β, dissociation with Triton X-100, immunoprecipitation with α-serum; lane 3: as in 2, but mitochondria were treated with 25 μg/ml trypsin prior to dissociation; lane 4: import of pre-β, SDS dissociation, β-specific antibody; lane 5: as in 4, but with trypsin treatment; lane 6: import of pre-β from translation of total yeast mRNA using SDS dissociation and β specific serum; lane 7: as in 6, but with trypsin treatment; lane 8: β precursor translated from T_7 transcript.

Precursors for α and β subunits were prepared by cloning the genes for the α and β subunits (*ATP1* and *ATP2*, respectively) downstream from T_7 RNA polymerase promoter. Cloning of the α and β genes has been described by Takeda et al. (1985, 1986). The T_7 expression vectors containing *ATP* 1 and 2 genes were used to prepare synthetic mRNAs for α and β subunits in a manner similar to that described by Horwich et al., (1985). This RNA was used to program synthesis of radiochemically pure α and β subunits in a reticulocyte lysate cell-free protein synthesis system.

Results of an import experiment using precursor to β are shown in Fig. 2. When β precursor was incubated with mitochondria, most of it was converted into mature β which was resistant to exogenously added protease, suggesting that the mature form was protected by the mitochondrial membranes (Fig. 2, lanes 4 and 5). In the same import experiment, if we dissociated mitochondria with non-ionic detergent and immunoprecipitated with antiserum to the α subunit, a fraction of the β subunit imported was co-immunoprecipitated (lanes 2 and 3). This fraction was significantly less than the total amount imported (lanes 4 and 5) but comparable to the amount imported when total yeast mRNA was used to direct protein synthesis (lanes 6 and 7).

From this experiment, we conclude that β subunit could be imported independently and could assemble with unlabeled α subunit already present in the mitochondria, so that it was immunoprecipitable with antiserum to the α subunit. We also found that β subunit was imported in excess of its capacity to form F_1.

These results were visualized better when we imported β precursor with and without precursor to α (Fig. 3). Separately added β and α precursors were imported, processed and immunoprecipitated with α-specific antiserum (lanes 1 and 2). In the

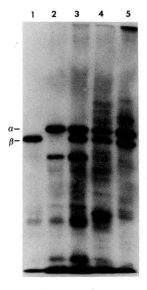

Fig. 3. Import and Assembly of α and β Precursors. α and β precursors transcribed and translated *in vitro* were incubated with isolated mitochondria either separately or together, dissociated with Triton X-100 and immunoprecipitated with α-specific antiserum. Lane 1: β subunit import; lane 2: α subunit import; lane 3: α and β subunit import; lane 4: import and assembly of F_1 subunits from translation of total yeast RNA, α immunoprecipitation; lane 5: mature F_1 subunits isolated from continuously labeled yeast cells.

case of the α subunit, a major degradation product of 40,000 MW was also detected. This peptide is a well-known proteolytic derivative of α (Ryrie and Gallagher, 1979). This product was observed even when import was conducted in the presence of serine protease inhibitors. When α and β precursors were incubated with mitochondria (lane 3), mature forms of both co-precipitated with α serum. These correspond in size to the F_1 subunits imported using total yeast mRNA to direct translation (lane 4) and with mature F_1 subunits from continuously labeled yeast cells (lane 5). This experiment indicated that α and β subunits were imported into mitochondria independently and that some of the α subunit imported *in vitro* was degraded inside the mitochondria.

TIME COURSE OF F_1 IMPORT

Using equimolar amounts of precursors to the α and β subunits made *in vitro*, we examined the rate of appearance of newly imported β subunit in isolated mitochondria. Precursors of the two largest subunits were incubated with mitochondria, and samples were withdrawn at increasing intervals. Synthesis and import was halted by inhibitors, and mitochondria were isolated by centrifugation. Trypsin treatment was used to remove precursors which adhered to the outside of the mitochondria. Samples were dissociated in SDS and separated on polyacrylamide gels, an autoradiogram of which is shown in Fig. 4. While the α subunit showed no increase in import after 10 minutes, the level of β subunit continued to rise throughout the 60

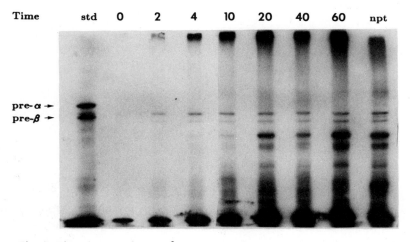

Fig. 4. Time Course of α and β Import. Equimolar amounts of precursor forms of α and β were incubated with isolated mitochondria, and at specific intervals aliquots of the reaction mix were removed and assayed for subunit import as described in the text. Times of import are indicated above each lane. Std: α and β precursors translated from their T_7 transcripts; npt: 60 min. time point without trypsin treatment.

minute incubation. These results reflect the results we obtained in measuring the rate of α and β subunits imported from a reticulocyte lysate programmed with total yeast RNA (Fig. 1). It appears that α is imported more rapidly than β, but that the intramitochondrial level of β increases gradually.

We noted that there was a dramatic increase in the level of the 40,000 MW fragment of α in the isolated mitochondria after 10 minutes of incubation. Our conclusion is that when α is imported in vast excess of the available β subunits, much of it is degraded by proteolysis. Hence, the mitochondria maintain a transient pool of excess α (and probably γ subunits) related to the fact that these subunits are imported more rapidly than β (Burns and Lewin, 1986). Unassembled α subunit is sensitive to proteolytic turnover. These conclusions have been substantiated by kinetic labeling experiments done *in vivo* at 15°C, in which mature α subunits appeared in mitochondria more rapidly than β subunits (data not shown).

A COLD-SENSITIVE MUTATION IN *oli1*

Our second set of experiments involved the interaction of F_o subunits with proteins encoded in the nucleus. We have recently described (Hefta et al., 1987) a mitochondrial mutant called 990 which has a point mutation mapping to the *oli1* gene of yeast mitochondrial DNA. *Oli1* encodes subunit 9 of the ATPase complex, the DCCD-binding proteolipid (Macino and Tzagoloff, 1979; Hensgens et al., 1979). The phenotype of this mutation is: (1) cold sensitive growth on non-fermentable carbon sources, such as glycerol; (2) low ATPase activity; (3) resistance to the antibiotic oligomycin at 3 μg/ml; and (4) altered electrophoretic mobility of subunit 9 on SDS-polyacrylamide gels.

REVERTANTS OF 990

In order to study the interaction of F_1 subunits with F_o, we selected revertants of 990 following mutagenesis of this strain with manganese chloride. We obtained 7 true revertants, that is, back-mutations at the original site. We also obtained two strains which appeared to contain nuclear suppressor mutations based on their segregation patterns in meiosis (2:2 as expected for Mendelian genes, rather than 4:0 as expected for mitochondrial genes). We obtained one mutation (R71) which behaved like a mitochondrial revertant at a second site. Analysis of the ATPase activities of mitochondria isolated from these strains is presented in the Table. Mutant strain 990 exhibited roughly one quarter of the ATPase activity of the wild-type strain, but a true revertant (R74) and a strain harboring a second mitochondrial mutation (R71) exhibited normal levels of activity. ATPase activities in the two nuclear suppressor strains (R61 and R85) were only slightly higher than in the original 990 mutation. The strains exhibited converse behavior with respect to oligomycin resistance: Strain 990 was four times more resistant to oligomycin than the wild-type strain and the mitochondrial revertants. Strains R61 and R85, the nuclear revertants, exhibited intermediate levels of resistance.

The migration of subunit 9 from several strains on SDS polyacrylamide gels is demonstrated in Fig. 5. Subunit 9 from *oli1* mutant 990 exhibits retarded electrophoretic mobility relative to subunit 9 from wild-type yeast. The true revertants exemplified by strain R33 restore wild-type mobility. Strain R71 contains a mitochondrial reversion which is not a true back mutation (see below). Electrophoretic mobility of subunit 9 from this strain is more rapid than that from 990 and, indeed, from that of the wild-type.

The mutation which underlies this pleiotropic phenotype of 990 is a C to T transition which leads to a change from an alanine to a valine residue at position 22 of the protein (Fig. 6). In the true revertants, the sequence of the gene is returned to that of the wild-type parent. In strain R71, however, the T present in codon 22 of 990 is altered to a G. This converts the valine codon to a glycine codon. It appears, therefore, that a small side-chain amino acid is required at this position for full function of the ATPase. The enzyme with glycine in this position is nearly as active as the wild-type enzyme. Position 22 falls between two glycines in yeast subunit 9, and glycines at those positions are conserved in a variety of organisms (Willson and Nagley, 1987). We suspect that a turn or other flexible region is demanded at this site in the protein.

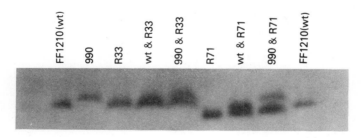

Fig. 5. [^{14}C]DCCD Labeled Proteolipid from 990 and Revertants.

NUCLEAR SUPPRESSORS

Our nuclear suppressors of 990 (R61 and R85) are recessive in phenotype. 990 diploids heterozygous for the nuclear mutation exhibit the cold-sensitive growth on glycerol just like the haploid strain without the suppressor. In addition, both nuclear suppressors when separated from the 990 mutation in the haploid state are cold-sensitive for growth on glycerol. The two suppressors we have isolated probably fall in independent genes: diploids are cold-sensitive. The mutations in R61 and R85 may affect ATPase subunits. When the nuclear genes are transferred from the 990 mitochondrial background to a strain containing another oligomycin-resistant form of the *oli*1 gene, the strains exhibit deficiency of ATPase activity which segregates with the nuclear suppressor allele in meiosis.

DISCUSSION

Our experiments can lead to several conclusions about the formation of the mitochondrial ATPase complex: First, the formation of new F_1 may be regulated by the availability of one subunit. Since β is imported more slowly than α or γ, its transport into mitochondria may be a rate-limiting step in ATPase synthesis. Second, excess α subunit within the mitochondria appears to be degraded in the absence of F_1 assembly. Third, nuclear mutations selected on their ability to suppress a conditional mutation in *oli1*, may define genes in other ATPase subunits or in genes whose products are needed for F_1F_0 assembly.

Why should the largest F_1 subunits be transported at different rates? One explanation is that they react with a different receptor-molecule on the mitochondrial surface. No one, as yet, has identified a receptor for mitochondrial targeting peptides, if one exists. Although the pre-sequences of mitochondrial proteins are not identical,

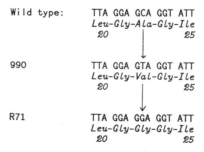

Fig. 6. Nucleotide and Amino Acid Changes in 990 and R71.

they do share some common features (Von Heijne, 1986) suggesting surface-active properties. At this stage we feel it is premature to guess whether there is more than one receptor for imported ATPase subunits. Our results could also be explained by a single receptor with differing affinities for α and β precursors.

The aim of our genetic experiments is to identify proteins, presumably F_1 subunits, which interact with F_0 subunits directly. While it is too early to tell if we have succeeded, we have established a _functional_ interaction between the products of our suppressor genes and the product of *oli1*, namely, subunit 9. When present in the same cell, these lesions result in a functional ATPase. When present in different cells, they each lead to a defective enzyme. We hope that this line of research will define surfaces of interaction between the F_1 and F_0 domains.

ACKNOWLEDGMENTS

This work was supported by NIH grant number GM 29387-06.

REFERENCES

Burns, D.J. and Lewin, A.S. (1986) J. Biol. Chem. 261, 12066-12073.

Hefta, L.J.F., Lewin, A.S., Daignan-Fornier, B. and Bolotin-Fukuhara, M. (1987) Mol. Gen. Genet. 207, 106-113.

Hensgens, L.A.M., Grivell, L.A., Borst, P. and Bos, J.L. (1979) Proc. Natl. Acad. Sci. USA 76, 1663-1667.

Horwich, A.L., Kalousek, F., Mellman, I. and Rosenberg, L.E. (1985) EMBO J. 4, 1129-1135.

Lewin, A.S. and Norman, D.K. (1983) J. Biol. Chem. 258, 6750-6755.

Macino, G. and Tzagoloff, A. (1979) J. Biol. Chem. 254, 4617-4623.

Neupert, W. and Schatz, G. (1981) Trends Biochem. Sci. 6, 1-4.

Ryrie, I.J. and Gallagher, A. (1979) Biochim. Biophys. Acta 545, 1-14.

Schatz, G. and Butow, R.A. (1983) Cell 32, 316-318.

Takeda, M., Vassarotti, A. and Douglas, M.G. (1985) J. Biol. Chem. 260, 15458-15465.

Takeda, M., Chen, W., Saltzgaber, J. and Douglas, M.G. (1986) J. Biol. Chem. 261, 15126-15133.

von Heijne, G. (1986) EMBO J. 5, 1335-1342.

Willson, T.A. and Nagley, P. (1987) Eur. J. Biochem., in press.

ASSEMBLY OF YEAST MITOCHONDRIAL ATP SYNTHASE INCORPORATING AN IMPORTED VERSION OF AN Fo SECTOR SUBUNIT NORMALLY ENCODED WITHIN THE ORGANELLE

Leigh B. Farrell, Debra Nero, Sybella Meltzer, Giovanna Braidotti,
Rodney J. Devenish and Phillip Nagley

Centre for Molecular Biology and Medicine and Department
of Biochemistry, Monash University, Clayton, Victoria 3168, Australia

ABSTRACT

A new strategy for the controlled genetic manipulation of mitochondrial gene products is described, focusing on subunits of the mitochondrial ATP synthase (mtATPase) complex of *Saccharomyces cerevisiae*. Subunit 8 is an F_o-sector subunit 48 amino acids long and is normally a product of the mitochondrial *aap1* gene. An imported version of subunit 8 has been produced using an artificial nuclear gene. Import was mediated in yeast cells by use of an expression vector which directed the production of a precursor protein consisting of an N-terminal cleavable transit peptide, derived from the nuclearly encoded mtATPase subunit 9 from *Neurospora crassa*, fused to subunit 8. The imported version of subunit 8 assembles with the mitochondrially encoded subunits 6 and 9 in yeast to produce a functional F_o-sector. Parallel experiments are in progress with a version of yeast subunit 9 (76 amino acids long) specified by an artificial nuclear gene. These studies open the way to a directed mutagenesis approach to the analysis of mitochondrially encoded membrane-associated proteins of mitochondrial ATP synthase.

INTRODUCTION

In the yeast *Saccharomyces cerevisiae* the membrane-embedded F_o-sector of the mitochondrial proton-translocating ATP synthase (mtATPase) contains three subunits synthesized within the organelle (Linnane et al., 1985). These subunits, denoted 6, 8 and 9, are encoded respectively by the *oli2*, *aap1* and *oli1* genes in mtDNA, and are 259, 48 and 76 amino acids in length. Biochemical analysis of an extended series of mutants carrying mutations in these mitochondrial genes has provided insight into the

Molecular Structure, Function, and Assembly of the ATP Synthases
Edited by Sangkot Marzuki
Plenum Press, New York

assembly of the F_O-sector as a whole and into functionally significant regions of individual polypeptides.   Such protein domains are involved with the energy-transduction functions of the F_O-sector (proton channel and coupling to F_1) and its interaction with inhibitory drugs such as oligomycin (summarized by Linnane et al., 1985; Nagley & Linnane, 1987; see also papers by Marzuki et al., and Lukins et al., this volume).

Site-directed *in vitro* mutagenesis provides a powerful tool to create new mutations in order to probe structure-function relationships of proteins.   The contemporary technologies of molecular genetic manipulation of *S. cerevisiae* (Struhl, 1983) render this approach readily accessible for subunits of mtATPase encoded by nuclear genes in this organism (Vassarotti et al., 1987).   By contrast, corresponding manipulations of mitochondrially encoded proteins present considerable difficulties in the absence of a suitable transformation system for introducing DNA into the mitochondria of yeast cells and achieving its expression inside the organelle (Nagley et al., 1985). We therefore devised an approach for the controlled genetic manipulation of mitochondrially encoded proteins which is based on the expression of such proteins from artificial nuclear genes (Gearing et al., 1985). The expression strategy is designed (Nagley et al., 1985) so that the proteins will be imported into mitochondria from their cytosolic site of biosynthesis (Gearing and Nagley, 1986).   The usefulness of this strategy, particularly in concert with *in vitro* mutagenesis, depends upon the ability of the nuclearly encoded version of a mitochondrial gene product to assemble functionally into the mtATPase complex in place of its natural counterpart.   In this article, we summarize our recent findings that this can indeed be achieved with subunit 8.   We also outline the progress being made in parallel experiments using a nuclearly encoded version of subunit 9.

TOPOLOGICAL CONSIDERATIONS

The two proteolipid F_O-sector subunits 8 and 9 were initially chosen for this work on the basis of their relatively small size. The correspondingly small sizes of their genes simplify the task of nucleotide re-coding required for subsequent nuclear expression (see below). Both of these intensely hydrophobic subunits, when assembled into the mtATPase complex, are integral proteins of the inner mitochondrial membrane (Fig. 1). It has been proposed that the 48 amino acid length of subunit 8 includes one transmembrane segment (Velours et al., 1984).   From affinity labeling data, it was suggested (Velours and Guerin, 1986) that the positively charged C-terminus of subunit 8 lies on the matrix face of the inner mitochondrial membrane. By inference, the N-terminus is projected towards the intermembrane space (Fig. 1).

Subunit 9 is 76 amino acids long. It has two transmembrane stems separated by a hydrophilic loop which is suggested to face the F_1-sector (Sebald and Hoppe, 1981) and, therefore, in mtATPase to extend towards the matrix (Fig. 1). This loop presumably interacts with F_1 or other nuclearly encoded mtATPase subunits as suggested by the properties of yeast mutants affecting the loop region of subunit 9 (Willson and Nagley, 1987; Lukins et al., this volume).

Normally, these two subunits 8 and 9 are delivered to the inner membrane following their translation on mitochondrial ribosomes. In both cases, given the orientations shown in Fig. 1, the N-terminus of each protein crosses this membrane to face the intermembrane space. Let us consider what would have to happen in the case

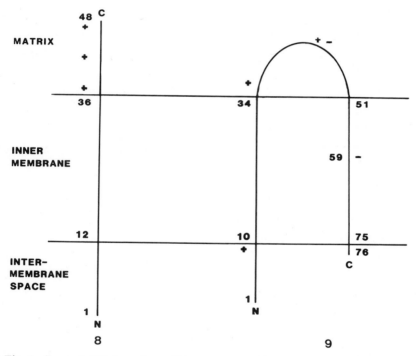

Fig. 1. Suggested Orientations of F_0-sector Subunits 8 and 9 Across the Inner Mitochondrial Membrane of *S. cerevisiae* (see text for literature citations). The N- and C-termini of the polypeptides and some amino acid residue numbers are indicated. The approximate positions of charged residues are indicated by + and − symbols. The tentative location of the boundaries of transmembrane domains were inferred from hydropathy plots of subunit 8 (Velours et al., 1984; our data not shown) and subunit 9 (Willson and Nagley, 1987). The membrane-external portions of the polypeptides are drawn as projecting from the membrane merely for convenience; this representation does not imply a particular configuration for these domains with respect to the surface of the membrane. The proteins form specific aggregates amongst themselves and with other F_0-sector subunits, and possibly also with nuclearly-encoded F_1-sector subunits that join to the membrane at the matrix side.

of imported versions of these proteins, where the correct orientation would need to be adopted following their import across the membrane. The proposed scheme of events shown in Fig. 2 is based on the following general features of protein import into mitochondria (reviewed by Hay et al., 1984; Hurt and van Loon, 1986; Douglas et al., 1986). The N-terminus of an imported protein is pulled in towards the matrix trailing the positively charged N-terminal leader sequence. This leader acts as a transit peptide targeting the protein to mitochondria and is generally cleaved away from its passenger protein by a chelator-sensitive proteolytic enzyme located in the matrix space (known as matrix protease). The imported protein then is further sorted to its submitochondrial destination.

The proposed events are illustrated in Fig. 2 in specific terms concerning the imported version of subunit 8. This scheme incorporates the use of the transit peptide from the precursor of the nuclearly encoded mtATPase subunit 9 of *Neurospora crassa* (Viebrock et al., 1982; Schmidt et al., 1983b). The fusion protein, consisting of this N-terminal leader joined to subunit 8, is denoted N9/Y8 (Fig. 2, A). This particular transit peptide was chosen for two reasons. First, the natural precursor of *N. crassa* subunit 9 is efficiently imported by mitochondria of *S. cerevisiae* where it is correctly processed (Schmidt et al., 1983a). Second, the protein this leader delivers naturally to mitochondria is an authentic mtATPase subunit 9, which must undergo the appropriate

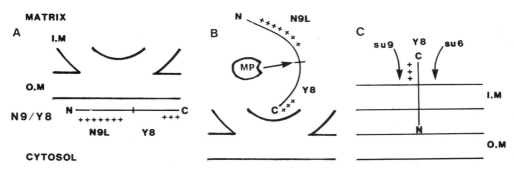

Fig. 2. Scheme for Import of Subunit 8 into Mitochondria and Assembly into F_o-sector of mtATPase (see text for details and literature citations). A. The precursor N9/Y8 binds to isolated mitochondria, probably by a receptor molecule. The transit peptide from *N. crassa* subunit 9 precursor is denoted N9L; the remainder, lying to the right of the vertical bar, is substantially yeast subunit 8 and is denoted Y8. B. Import of N9/Y8 precursor (N–terminus first, occurring in several steps) and its cleavage by matrix protease (MP) to release Y8, which probably remains closely associated with the membrane because of its intense hydrophobicity. C. Reorientation of Y8 into the membrane, to give its final orientation (N–terminus facing out; cf. Fig. 1). Y8 interacts sequentially with subunits 9 and 6 to form an active F_o-sector, as would the natural mitochondrially encoded subunit 8 protein. Other indications: O.M, outer membrane; I.M, inner membrane (these are considered to be fused at the locations of protein import into mitochondria).

reorientations needed to implant it in the inner mitochondrial membrane in the orientation shown for its yeast equivalent in Fig. 1. Hartl et al. (1986) have suggested that this occurs by the import of the complete precursor into the matrix space although the subunit 9 precursor appears to retain some association with the membrane (Zwizinski and Neupert, 1983). Cleavage of the transit peptide by matrix protease in two steps (Schmidt et al., 1984) is followed by re-orientation of the mature protein into the membrane from within the organelle.

The anticipated behaviour of subunit 8, following import and cleavage from its transit peptide (Fig. 2, B), is that it would become available for assembly with the other F_o-sector subunits 6 and 9 (Fig. 2, C). The predicted sequential interactions first with subunit 9 and then with subunit 6 (Linnane et al., 1985) are those in which subunit 8 would participate as a mitochondrially synthesized protein, to lead to the assembly of a functional F_o-sector coupled to the F_1-sector. The experimental validation of the overall proposition represented by the events depicted in Fig. 2 is summarized in the following section.

FUNCTIONAL REPLACEMENT OF SUBUNIT 8 WITH AN IMPORTED VERSION EXPRESSED FROM AN ARTIFICIAL NUCLEAR GENE

At the outset, the *aap1* gene was redesigned for optimal compatibility with codon usage in the yeast nucleo-cytosolic system (Gearing et al., 1985). One codon was changed to accommodate a specific codon dictionary difference between yeast mtDNA and nuclear DNA; thirty other codons were adjusted to provide codons most frequently used in highly expressed yeast genes. The new gene, *NAP1*, was built by total chemical synthesis (Gearing et al., 1985), and could be expressed by *in vitro* transcription/translation as a radiolabelled protein showing properties of authentic subunit 8 (Nagley et al., 1985). A DNA segment derived from a cDNA clone of *N. crassa* (Viebrock et al., 1982), encoding the 66 amino acid transit peptide and the first 5 amino acids of mature subunit 9, was fused to the *NAP1* gene. The gene fusion was made in such a way that in the encoded N9/Y8 protein there were introduced an extra 2 serine residues just upstream of the subunit 8 sequences (Gearing and Nagley, 1986). This N9/Y8 precursor protein, produced by *in vitro* transcription/translation, was taken up by energized isolated yeast mitochondria and was processed at the matrix protease cleavage site of the *N. crassa* subunit 9 precursor, thus generating an imported subunit 8 with an N-terminal extension of 7 amino acids (Gearing and Nagley, 1986).

For tests of the ability of the imported subunit 8 to be correctly assembled in the mtATPase complex, the N9/Y8 protein was expressed *in vivo*, using cells of *S.*

cerevisiae with defective *aap1* genes unable to synthesize functional subunit 8 inside mitochondria. For this purpose an expression vector pLF1 was constructed (Nagley et al., 1987), which replicates as a multicopy plasmid in the yeast nucleus and which carriesthe nuclear *ADE1* gene to allow selection of transformants. The *aap1* mutant hosts carry the auxotropic *ade1* marker, being all derived from the parent rho^+ strain J69-1B (Macreadie et al., 1983). The expression site of pLF1 is within a transcriptional unit (Devenish et al., 1987) based on the promoter and transcription termination signals of the highly expressed yeast phosphoglycerokinase gene.

When *aap1 mit⁻* mutants, unable to grow on ethanol as carbon source because of defects in the mtATPase complex (Macreadie et al., 1983; Linnane et al., 1985), were transformed with pLF1-N9/Y8 which actively expresses the N9/Y8 protein, growth on ethanol was restored in each case tested (Table I). The mutant hosts included strains with severely truncated subunit 8 polypeptides 2 or 18 amino acids in length. Further genetic analyses (Nagley et al. 1987) confirmed two critical points. First, the rescued phenotype for each mutant host depended on the presence of the pLF1-N9/Y8 plasmid. Second, the host cells still contained defective *aap1* genes in otherwise intact mitochondrial genomes. Additionally, when the vector pLF1-Y8 (i.e. pLF1 carrying a *NAP1* gene, encoding subunit 8 not fused to a transit peptide) was introduced into the *aap1 mit⁻* mutant hosts, no rescue of their defective ethanol growth phenotype was observed (data not shown).

The inference is made that the import of subunit 8 is responsible for rescue of the growth defect, presumably by genetic reconstitution of the mtATPase complex. The lack of gross temperature-sensitivity of the rescued phenotype (Table I) is suggestive of an efficient functional assembly of imported subunit 8 into the mtATPase complex. This has been directly shown (Nagley et al., 1987) in the case of the *aap1 mit⁻* mutant strain M31 transformed with pLF1-N9/Y8. Analysis of immunoprecipitates of mtATPase using a monoclonal antibody specific for the β-subunit of the F_1-sector (Hadikusumo et al., 1984) showed the mitochondrially synthesized subunit 8 to have been replaced by a very abundant cytosolically synthesized imported polypeptide of slightly greater length (corresponding to subunit 8 with an N-terminal extension of 7 amino acids).

Biochemical analyses of mtATPase performance in the case of M31 transformed with PLF1-N9/Y8 showed that the genetically reconstituted mtATPase was generally comparable in its activities to that of the rho^+ strain J69-1B. Nevertheless there appeared to be some minor perturbation of proton channel function or coupling efficiency in M31 expressing N9/Y8, manifested in mitochondria by slightly reduced oligomycin-sensitivity of mtATPase activity, and by lowered ATP-P$_i$

Table I. Growth on Ethanol Medium of *aap1* Mutants and Transformants Expressing the N9/Y8 Protein[a]

Strain	Length of *aap1* product (amino acids)[b]	Transformed with PLF1-N9/Y8	Relative growth rate 18°C	28°C	36°C
J69-1B (wild-type)	48	no	+++	+++	+++
		yes	+++	+++	+++
M31	2	no	-	-	-
		yes	+++	+++	++
M68-5	18	no	-	-	-
		yes	+++	+++	+++
M26-10	18 (1 subst)	no	-	-	-
		yes	+++	+++	+++

[a] Growth was tested by dropping aliquots of aqueous suspensions of cells onto YEPE-agar plates (Linnane et al., 1975) containing ethanol as carbon source. The extent of growth of patches was scored after 5 days, relative to that of the parent *rho*[+] (defined as +++ at each temperature). Growth of all cells on YEPD-agar (containing glucose) was comparable for each temperature (not shown). Details of mutations in the *aap1* gene and the predicted subunit 8 derivatives are as reported by Macreadie et al. (1983). [b] Number of substitutions are indicated as subst.

exchange activity. It has not been ascertained whether this is due to the extra 7 amino acids attached to the imported subunit 8 or to the association of supernumerary subunit 8 polypeptides with the mtATPase complex arising from over-production from the multicopy expression vector (Nagley et al., 1987). In order to check this point, the construction of single-copy vectors or chromosomally integrated DNA segments expressing N9/Y8 is being undertaken in an effort to reduce the amount of subunit 8 entering mitochondria.

EXPRESSION OF SUBUNIT 9 OUTSIDE MITOCHONDRIA USING AN ARTIFICIAL NUCLEAR GENE

In parallel with the above studies on subunit 8, we have designed and constructed by total chemical synthesis a new subunit 9 gene (*NAP9*) that is compatible with the nucleo-cytosolic expression system of yeast (Farrell et al., 1987). The sequence of the *NAP9* gene includes preferred codons used by highly expressed yeast genes. *NAP9* also features a series of 14 unique restriction enzyme cleavage sites embedded within the coding region for subunit 9, more or less evenly distributed throughout the gene. This feature facilitates subsequent genetic manipulation of subunit 9. For example, it will allow a cassette type of mutagenesis whereby introduction of a barrage of mutations affecting discrete domains of the protein can be carried out.

The *NAP9* gene has been expressed by *in vitro* transcription/ translation into a radiolabelled protein which shows the same mobility on polyacrylamide gel electrophoresis and equivalent solubility in chloroform-methanol as authentic subunit 9 (Farrell et al., 1987). A putative precursor N9/Y9 bearing the transit peptide and the first five amino acids of *N. crassa* subunit 9 fused, via two serines, to yeast subunit 9, has been expressed *in vitro* (this construct mirrors N9/Y8). Preliminary experiments indicate that N9/Y9 is taken up by energized isolated yeast mitochondria to generate an imported version of yeast subunit 9 (L.B. Farrell and P. Nagley, unpublished data). A pLF1 vector carrying DNA encoding N9/Y9 is under construction and will be used to transform *oli1 mit⁻* mutants (Ooi et al., 1985) unable to synthesize a functional subunit 9 inside mitochondria.

CONCLUSIONS

The successful replacement of mitochondrial subunit 8 by a nuclearly coded version will now directly lead to *in vitro* mutagenesis experiments whereby the structure-function relationships of subunit 8 can be probed in more depth than is possible using the currently available naturally occurring *aap1* mutants. In particular, it should be possible to produce mutants which will allow us to evaluate whether subunit 8 plays a direct role in proton conductance or energy coupling over and above its known role (Linnane et al., 1985) in assembly of the F_o-sector. If yeast subunit 9 can be successfully imported and assembled into mtATPase as has been accomplished with subunit 8, further powerful tools to dissecting the proton channel itself will become available. Finally, if the subunit 6 gene can eventually be reprogrammed for nuclear expression and an imported version of subunit 6 functionally assembled into the mtATPase complex, the assembly of this complex in the absence of any direct contribution from mitochondrial genes becomes a real possibility.

REFERENCES

Devenish, R.J., Maxwell, R.J., Beilharz, M.W., Nagley, P. and Linnane,A.W. (1987) Develop. Biol. Standard. 67, 185-199.

Douglas, M.G., McCammon, M.T. and Vassarotti, A. (1986) Microbiol. Rev. 50, 166-178.

Farrell, L.B., Gearing, D.P. and Nagley, P. (1987) manuscript in preparation.

Gearing, D.P., McMullen and Nagley, P. (1985) Biochem. Int. 10, 907-915.

Gearing, D.P. and Nagley, P. (1986) Eur. J. Biochem. 5, 3651-3655.

Hadikusumo, R.G., Hertzog, P.J. and Marzuki, S. (1984) Biochim. Biophys. Acta 765, 258-267.

Hartl, F.U., Schmidt, B., Wachter, E., Weiss, H. and Neupert, W. (1986) Cell 47, 939-951.

Hay, R., Bohni, P. and Gasser, A. (1984) Biochim. Biophys. Acta 779, 65-87.

Hurt, E.C. and van Loon, A.P.G.M. (1986) Trends Biochem. Sci. 11, 204-207.

Linnane, A.W., Kellerman, G.M. and Lukins, H.B. (1975) in Techniques of Biochemical and Biophysical Morphology, Vol. 2 (Glick, D. and Rosenbaum, R., Eds.) pp 1-98, John Wiley and Sons, New York.

Linnane, A.W., Lukins, H.B., Nagley, P., Marzuki, S., Hadikusumo, R.G., Jean-Francois, M.J.B., John, U.P., Ooi, B.G., Watkins, L., Willson, T.A., Wright, J. and Meltzer, S. (1985) in Achievements and Perspectives of Mitochondrial Research, Vol. I, Bioenergetics (Quagliariello, E., Slater, E.C., Palmieri, F., Saccone, C. and Kroon, A.M., Eds.) pp 211^222, Elsevier Science Publishers, Amsterdam.

Macreadie, I.G., Novitski, C.E., Maxwell, R.J., John, U., Ooi, B.G., McMullen, G.L., Lukins, H.B., Linnane, A.W. and Nagley, P. (1983) Nucl. Acids Res. 11, 4435-4451.

Nagley, P., Farrell, L.B. Gearing, D.P., Nero, D., Meltzer, S. and Devenish, R.J. (1987) submitted for publication.

Nagley, P. and Linnane, A.W. (1987) In Structure and Function of Energy Transducing Systems (Ozawa, T., and Papa, S., Eds.) Japan Scientific Societies Press, Tokyo, in press.

Nagley, P., Willson, T.A., Tymms, M.J., Devenish, R.J. and Gearing, D.P. (1985) in Achievements and Perspectives of Mitochondrial Research, Vol. II, Biogenesis (Quagliariello, E., Slater, E.C., Palmieri, F., Saccone, C. and Kroon, A. M., Eds.) pp 405-414, Elsevier Science Publishers, Amsterdam.

Ooi, B.G., McMullen, G.L., Linnane, A.W., Nagley, P. and Novitski, C.E. (1985) Nucl. Acids Res. 13, 1327-1339.

Schmidt, B., Hennig, B., Kohler, H. and Neupert, W. (1983a) J. Biol.Chem. 258, 4687-4689.

Schmidt, B., Hennig, B., Zimmerman, R. and Neupert, W. (1983b) J. Cell Biol. 96, 248-255.

Schmidt, B., Wachter, E., Sebald, W. and Neupert, W. (1984) Eur. J. Biochem. 144, 581-588.

Sebald, W. and Hoppe, J. (1981) Curr. Top. Bioenerg. 12, 1-64.

Struhl, K. (1983) Nature 305, 391-397.

Vassarotti, A., Chen, W.J., Smagula, C. and Douglas, M.G. (1987) J. Biol.Chem. 262, 411-418.

Velours, J., Esparza, M., Hoppe, J., Sebald, W. and Guerin, B. (1984) EMBO J. 3, 207-212.

Velours, J. and Guerin, B. (1986) Biochem. Biophys. Res. Commun. 138, 78-86.

Viebrock, A., Perz, A. and Sebald, W. (1982) EMBO J. 1, 565-571.

Willson, T.A. and Nagley, P. (1987) Eur. J. Biochem. in press.

Zwizinski, C. and Neupert, W. (1983) J. Biol. Chem. 258, 13340-13346.

STRUCTURE AND FUNCTION OF ATP SYNTHASE

STRUCTURE OF THE *ESCHERICHIA COLI* ATP SYNTHASE FROM

ELECTRON MICROSCOPY STUDIES

Uwe Lucken, Edward P. Gogol and Roderick A. Capaldi

Institute of Molecular Biology, University of Oregon
Eugene, Oregon 97403

INTRODUCTION

ATP synthesis in both prokaryotes and eukaryotes is carried out by a large membrane-bound complex made up of two functionally and structurally distinct parts. The catalytic part, the F_1, is extrinsic to the membrane bilayer, and is attached to the integral membrane assembly, the F_0, which provides the proton channel (Senior and Wise, 1983; Walker et al., 1984; Amzel and Pedersen, 1983).

The ATP synthases from bacteria, chloroplasts and mitochondria are very similar with several of the subunits highly conserved through evolution (Walker et al., 1983, 1985). The bacterial enzyme, as represented by the ATP synthase of *Escherichia coli* (ECF_1F_0), is the simplest structurally and best characterized of these. The F_1 part of the *E. coli* enzyme (ECF_1) is made up of five different subunits (α, β, γ, δ and ϵ) in the ratio 3:3:1:1:1. The F_0 part (ECF_0) contains three subunits (a, b and c) in the ratio 1:2:10-20 (Senior and Wise, 1983; Foster and Fillingame, 1979).

We are examining the three-dimensional structure of ECF_1F_0 by electron microscopy of both negatively stained and unstained frozen hydrated specimens. This paper reports on some of the key features that have been established so far, including the size and shape of the complex and the relationship of its components.

STUDIES OF NEGATIVELY STAINED TWO-DIMENSIONAL CRYSTALS OF ECF_1

Macromolecular assemblies like the ECF_1 and ECF_1F_0 can be visualized and characterized structurally by electron microscopy.   Use of low electron doses

Molecular Structure, Function, and Assembly of the ATP Synthases
Edited by Sangkot Marzuki
Plenum Press, New York

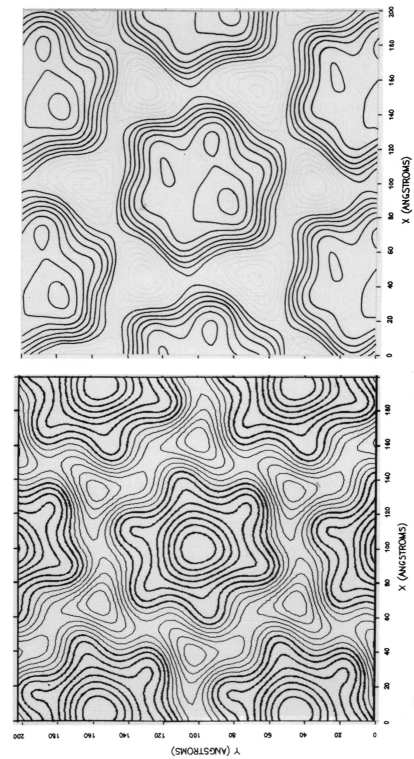

Fig. 1. Contour Projection Map of Two-Dimensional Crystals of ECF_1 Stained with Phophotungstate Alone (left) and Phosphotungstate plus Uranyl Acetate (right).

minimizes radiation damage to the specimens, but yields noisy images whose direct interpretation is difficult. The signal-to-noise ratio in the images, and thus the reliability of observed features, can be dramatically improved by averaging images of many molecules. Crystalline specimens, in which the spatial relationships among all molecules are defined, provide the most direct way to do this.

Two-dimensional arrays of F_1 from a thermophilic bacterium have been reported previously by Wakabayashi et al. (1977). Electron micrographs of these crystals stained with uranyl acetate were pseudo-optically filtered, partially averaging molecules in the arrays. The F_1 molecules appeared roughly hexagonal. Recently we have obtained two-dimensional crystals of ECF_1 in specimens of ECF_1F_o reconstituted into lipid vesicles by negatively staining with phosphotungstic acid (PTA). The PTA (at pH 7.0) dissociates a fraction of ECF_1 from the membranes and induces the formation of arrays on the electron microscope grids. These crystals exhibit p6 symmetry to a resolution of approximately 25 Å; images have been analyzed by Fourier methods, averaging the information from each molecule in the arrays. The resulting projection map, displayed as a density contour plot (Fig. 1A), shows six peripheral lobes around a stain-excluding central region. Three-dimensional information has been collected by recording micrographs of crystals tilted at angles spanning the range of 0-60°; a full reconstruction of this data is now underway, and will yield the outline of the ECF_1 complex.

The crystals of ECF_1 have also been examined by a double staining technique, in which ECF_1F_o is treated first with PTA on a thin layer of mica, dried, carbon shadowed, and then restained with uranyl acetate. Greater deviations from six-fold symmetry are seen with these samples. The resulting projection map (Fig. 1B) shows an asymmetric location for the central density in the structure, suggesting that the cationic uranyl acetate better penetrates the central region of the ECF_1 than the larger anionic PTA.

CRYOELECTRON MICROSCOPY OF ECF_1

The structure of ECF_1 and ECF_1F_o is also being examined in an unstained state by cryoelectron microscopy. Specimens are rapidly frozen into thin layers of amorphous ice directly from buffers in which they are fully active, with no additional chemical treatments. In the absence of stain, contrast in the image recorded under low dose conditions is that between protein, lipid and water (as ice) (Fig. 2). To date we have focused our efforts on examination of single particles of ECF_1 and ECF_1F_o. The rapid freezing of samples offers an opportunity to trap molecules in different functional states for structure determination. Also, it is likely that more subunit specific antibodies can be bound to single particles for topology studies than could be bound to crystalline preparations because of steric constraints in closely packed arrays.

Analysis of images of randomly-oriented single particles requires the use of methods to classify the variety of projections of the molecules, and to accurately align images within each category for averaging. Correlation and correspondence-based methods have been developed for this purpose (Frank et al., 1981; Van Heel, 1984), and have been applied to the study of negatively stained beef heart F_1 (Boekema et al., 1986). The projection structure of the uranyl acetate-stained F_1 is very similar to that obtained from the double-stain procedure with ECF_1 crystals (Fig. 1B). We have applied similar alignment and classification methods to the analysis of images of unstained ECF_1 (visualized by cryoelectron microscopy). The two most common classes of images of ECF_1 are represented by the gray-level contour plots in Fig. 3.

The first of these averaged images (Fig. 3A) is similar to the projections of the crystalline sample (Fig. 1B) and also to the results of single particle analysis of F_1 in stain (Boekema et al., 1986). Both the single particle data and the analysis of two-dimensional crystals show a hexagonally shaped molecule with dimensions of 80 x 85 Å in this one projection. There are six peripheral masses, an asymmetrically located seventh mass in the interior as well as a central cavity or cleft. The projection shown in Fig. 3B shows a bilobed structure consisting of two elongated densities separated by a region of lower density.

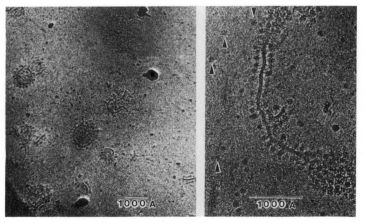

Fig. 2. Electron Micrographs of ECF_1F_0 in Membranes as Visualized by Cryoelectron Microscopy of Specimens Embedded in a Thin Layer of Ice.

The two projections in Fig. 3 have been related by comparing pairs of images of the single molecules rotated by 90°. This has been accomplished by recording micrographs at tilt angles of +45 and -45. Hexagonal projections were selected from both first and second exposures and the same molecules in the corresponding tilt (90° from the first) were then located. Two such tilt pairs are shown in Fig. 4. Each rotates from the hexagonal to the bilobed projection, indicating that the bilobed structure is the side view of the hexagonal projection.

CRYOELECTRON MICROSCOPY OF ECF_1F_0 IN MEMBRANES

The entire ECF_1F_0 complex has been examined after reconstitution with lipid into vesicular and sheet-like membranes.

The hexagonal projection of ECF_1 described above (Fig. 1B and 3A) is the predominant view of the intact ECF_1F_0 reconstituted into vesicles and viewed normal to the membrane (i.e. from the top). This projection is not seen when ECF_1F_0

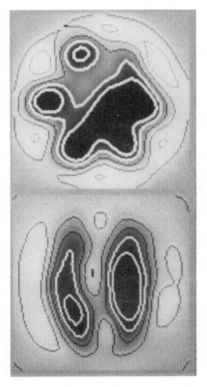

Fig. 3. Two Major Classes of Projections of ECF_1 Molecules Averaged from Electron Micrographs of Unstained Samples.

111

molecules are viewed from the side, either at vesicle edges or in sheets viewed edge on. Fig. 5 shows major classes of side views of ECF_1F_o. The ECF_1 appear bilobed in some of these projections with a cleft normal to the membrane. Others show a more solid appearance. The cleft seen in side views is not as distinct as in the isolated ECF_1 molecule, possibly because of some intercalation of the subunits of F_o (subunit *b*?).

The most significant feature of the ECF_1F_o molecules viewed from the side is the presence of a stalk separating the F_1 and F_o. This stalk is approximately 50 Å long and 20 Å wide, the width being approximately the size of four or five closely packed α-helices. A separation of F_1 from the membrane by as much as 50 Å has been seen in micrographs of negatively stained bacterial, mitochondrial and chloroplast membranes (Fernandez-Moran et al., 1964; Telford et al., 1984) but has often been discounted as an artifact of the staining procedures employed (Sjostrand et al., 1964; Wainio, 1985).

COUPLING MECHANISMS

Biochemical studies have shown that ECF_1 contains the active sites on β subunits, while the F_o makes up the proton channel, probably formed by both the *a* and *c* subunits (Senior and Wise, 1983; Walker et al., 1984). The stalk must serve to couple proton translocation across the membrane to hydrolysis and synthesis of ATP

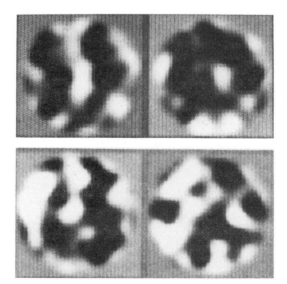

Fig. 4. Pairs of Images of ECF_1 in which the Hexagonal View (Upper Panel) is Visualized in a Second Micrograph Taken After a 90° Rotation of the Specimen (Lower Panel).

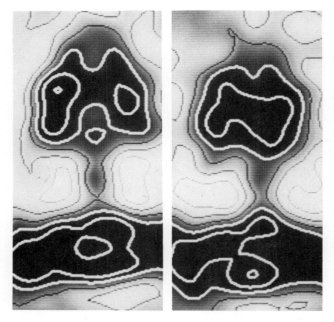

Fig. 5. Two Major Classes of Side Views of ECF_1F_o Averaged from Membranes Embedded in a Thin Layer of Ice.

in ECF_1. Coupling could be by proton transfer through the F_o and stalk, a distance in excess of 100 Å. This seems unlikely, in that there is no obvious way to shield(vectorial) protons from diffusing into the aqueous phase while being translocated through the stalk part. A more plausible coupling mechanism is that proton translocation induces a conformational change in the stalk, possibly a twisting of the helical elements which is then transmitted to the catalytic sites in the ECF_1. Clear evidence that the ECF_1 and F_o parts are conformationally coupled is provided by the studies of Penefsky (1985) who showed that DCCD bound in F_o altered the affinity for nucleotides in the active sites on ECF_1.

ACKNOWLEDGMENTS

The work was supported by National Institutes of Health grant HL 24526. UL thanks the German Academic Exchange Program for the NATO Fellowship 300-402-503-6.

REFERENCES

Amzel, L.M. and Pederson, P.L. (1983) Ann. Rev. Biochem. 52, 801-824.

Boekema, E.J., Berden, J.A. and Van Heel, M.G. (1986) Biochim. Biophys. Acta 851, 353-360.

Fernandez-Moran, H., Odat, T., Blair, P.V. and Green, D.E. (1964) J. Cell. Biol. 22, 63-100.

Foster, D.L. and Fillingame, R.H. (1979) J. Biol. Chem. 254, 8230-8236.

Frank, J., Verschoor, A. and Boublik, M. (1981) Science 214, 1353-1355.

Penefsky, H.S. (1985) Proc. Natl. Acad. Sci. USA 82, 1589-1593.

Saraste, M. and Tybulewicz, V.L.J. (1985) J. Mol. Biol. 184, 677-701.

Senior, A.E. and Wise, J.G. (1983) J. Membr. Biol. 73, 105-124.

Sjostrand, F.S., Andersson-Cedergren, E. and Karlsson, U. (1964) Nature 202, 1075-1078.

Telford, J.N., Langworthy, T.A. and Racker, E. (1984) J. Bioenerg. Biomemb. 16, 335-351.

Van Heel, M. (1984) Ultramicroscopy 13, 165-184.

Wainio, W.W. (1985) J. Ultrastruct. Res. 93, 138-143.

Wakabayashi, T., Kubota, M., Yoshida, M. and Kagawa, Y. (1977) J. Mol. Biol. 117, 515-519.

Walker, J.E., Fearnley, I.M., Gay, N.J., Gibson, N., Northrop, G.D., Powell, S.J., Runswick, M.J., Saraste, M. and Tybulewicz, V.L.J. (1985) J. Mol. Biol. 184, 677-701.

Walker, J.E., Saraste, M. and Gay, N.J. (1984) Biochim. Biophys. Acta 768, 164-200.

MONOCLONAL ANTIBODIES AS PROBES OF ASSEMBLY OF THE MITOCHONDRIAL ATP SYNTHASE

S. Marzuki, S.A. Noer, Herawati Sudoyo, Sybella Meltzer,
H.B. Lukins and A.W. Linnane

Department of Biochemistry and the Centre for Molecular
Biology and Medicine, Monash University, Clayton,
Victoria 3168, Australia

INTRODUCTION

Monoclonal antibodies to specific epitopes on the various subunits of the ATP synthase provide a powerful tool in the investigation of the molecular structure, function and assembly of the enzyme complex. In particular, monoclonal antibodies have been instrumental in the definition and the elucidation of the subunit composition of the enzyme complex (Hadikusumo et al., 1984, 1986; Moradi-Ameli and Godinot, 1983; Dunn et al., 1985; Joshi et al., 1985), the assembly pathway of its individual subunits (Marzuki et al., 1983; Linnane et al., 1985; Hadikusumo et al., 1988), and the subunit arrangement and the topography of the enzyme complex by immunoelectronmicroscopy (Lunsdorf et al., 1984). Molecular definition of epitopes recognized by monoclonal antibodies would allow the characterization of functional domains on the various subunits of the ATP synthase.

As part of our general strategy to elucidate the assembly process of the yeast mitochondrial H^+-ATPase - ATP synthase, a program was initiated several years ago to isolate a series of monoclonal antibodies with defined specificity to the various subunits of the yeast enzyme (Hadikusumo et al., 1984, 1986). This report examines the progress which have been achieved to date, in particular in the use of certain antibodies to define the assembly defects in a number of mit^- mutants with lesions in the structural genes of the membrane F_o subunits 6, 8 and 9 of the ATP synthase. The definition of these defects has allowed the assembly pathway of the F_o sector, and two novel nuclearly coded subunits, to be deduced. Subunit-specific antibodies to the human mitochondrial ATP synthase are being raised with a long term objective to define the enzyme assembly defects in patients with mitochondrial cytopathies.

MONOCLONAL ANTIBODIES TO THE YEAST MITOCHONDRIAL ATP SYNTHASE

A range of monoclonal antibodies to the yeast ATP synthase has been isolated in our laboratory. These monoclonal antibodies have been assigned into groups of common epitope specificity, based on their ability to compete against each other for binding to the enzyme complex in an Enzyme Linked ImmunoSorbent Assay (ELISA) (Hadikusumo et al., 1986). They have also been characterized in detail with regard to their subunit specificity, their effects on the ATPase activity, and their ability to precipitate the whole ATP synthase complex (Hadikusumo et al., 1984).

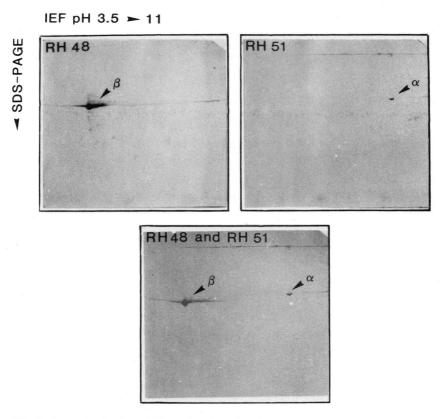

Fig. 1. Characterization of Monoclonal Antibodies to the Yeast ATP Synthase by Two Dimensional Gel Electrophoresis. F_1-ATPase was isolated from a wild-type strain of the yeast *Saccharomyces cerevisiae* (Beechey et al., 1975), and its protein subunits separated by two dimensional gel electrophoresis (first dimension - isoelectric focusing, second dimension - electrophoresis in SDS-polyacrylamide gel) as previously described (Marzuki and Linnane, 1982). After electrophoretic transfer to nitrocellulose filters, the filters were immunoblotted with various monoclonal antibodies. Shown are results obtained with monoclonal antibodies RH48, RH51 and their mixture.

Table I.   Properties of Monoclonal Antibodies to the Yeast ATP Synthase.

Epitopes	Clones	Inhibition of ATPase Activity (% inhibition by 100 mg IgG)	Ability to Immunoprecipitate Intact ATP Synthase
α	RH 72	-25	-
	51	-6	+
	71	8	-
	19	4	-
βI	RH 48	7	+
βII	57	90	-
	45	36	-
βIII	32	-3	-
βIV(?)	28	2	-
P25	RH 66		-
	41		-
	37		-

Monoclonal antibodies were assigned into groups of common epitope specificity by competitive ELISA (Hadikusumo et al., 1986). The subunit specificity of the antibodies was determined by two dimensional gel electrophoresis as in Fig. 1. The ability of the antibodies to inhibit the ATPase activity and to immunoprecipitate intact ATP synthase was assessed in a previous study (Hadikusumo et al., 1984).

One of the problems encountered in defining the subunit specificity of these monoclonal antibodies had been to distinguish between anti-α and anti-β-subunit antibodies, because of the close similarities in the molecular weights of these two subunits. The monoclonal antibodies, in particular those tentatively identified to be directed against the α and β subunits of the enzyme complex (Hadikusumo et al., 1984, 1986), therefore, have been recently re-screened by Western immunoblotting against the yeast F_1-ATPase, following the separation of its subunits by two dimensional gel electrophoresis (first dimension - isoelectric focusing, second dimension - polyacrylamide gel electrophoresis in the presence of sodium dodecyl sulphate; O'Farrel, 1975). Under this electrophoretic system, the β subunit is well separated from the α subunit which has a significantly higher isoelectric point (Fig. 1), allowing conclusive definition of the subunit specificity of the anti-α and anti-β monoclonal antibodies in our collection (Table I).

A group of four monoclonal antibodies previously thought to be directed against the β subunit (Hadikusumo et al., 1986) was found to react with the α subunit of the enzyme complex. Although these antibodies could be shown by competitive

ELISA to recognize a common epitope region, their antigenic determinants are most likely to be different, albeit presumably related and overlapping; the antibodies behave differently with regard to their ability to inhibit the ATPase activity and to immunoprecipitate intact ATP synthase complex.

The subunit specificity of the other monoclonal antibodies is as previously reported, with most of the antibodies directed against the β subunit. They are at least three epitope regions on this subunit. One of these epitopes is of a particular interest, as the two monoclonal antibodies which recognize this epitope region inhibit the ATPase activity, and thus is probably in the catalytic domain of the F_1-sector. The other epitopes recognized by the monoclonal antibodies are also of interest, as the binding of the antibodies to these epitopes results in the dissociation of a number the ATP synthase subunits, presumably due to conformational changes in the β subunit induced by the antibody binding (Hadikusumo et al., 1986).

Monoclonal antibodies have also been isolated to the ATP synthase subunits of pig (Moradi-Ameli and Godinot, 1983) and beef heart (Joshi et al., 1985) mitochondria, and of E. coli (Dunn et al., 1985). All of the anti-ATP synthase monoclonal antibodies which have been isolated to date, however, have been found to be directed against the hydrophilic subunits of the enzyme complex. While monoclonal antibodies to the hydrophobic subunits of the F_o sector would provide a powerful tool in the analysis of the structure and function of this sector, the inherent difficulty in obtaining such antibodies has been the high degree of hydrophobicity of the F_o-subunits.

To overcome this problem, a series of immunization protocols have been investigated. Five stable lines of myeloma-spleen cell hybrids, producing antibodies against the proteolipid subunit 9 of the yeast mitochondrial ATP synthase were successfully isolated, by immunizing mice with a proteolipid preparation in the presence of sodium dodecyl sulphate (Jean-Francois et al., 1988). All of the antibodies reacted with the yeast subunit 9 in an ELISA or in Western immunoblotting experiments. One of these monoclonal antibodies also recognized the ATP synthase subunit 8, indicating a shared epitopes. The antibodies did not react with the F_oF_1-ATPase, suggesting that their epitopes are shielded by other subunits of the enzyme complex. Interestingly, one of the anti-subunit 9 antibodies was found to react with the monomeric form of the ATP synthase subunit, but not to its oligomer, indicating that the epitope recognized by this antibody is involved in the interaction of subunit 9 molecules in the oligomer. The characterization of the epitopes recognized by the monoclonal antibodies would be of a particular interest with regard to the structural organization of the F_o proton channel.

THE DEFINITION OF THE SUBUNIT COMPOSITION OF THE YEAST MITOCHONDRIAL ATP SYNTHASE WITH MONOCLONAL ANTIBODIES: NOVEL POLYPEPTIDE SUBUNITS OF 18 kDa and 25 kDa

Information regarding the assembly sequence of the various subunits of a multimeric enzyme complex such as the ATP synthase can be deduced from the extent to which the assembly of the enzyme complex can proceed in the absence of the synthesis of one of its subunits. However, a major problem in the characterization of the defective mitochondrial ATP synthase in the mit^- mutants of yeast has been the instability of the incompletely assembled enzyme complex. In particular, the defective ATP synthase has been shown to be extremely cold-labile in the absence of subunit 9 (Orian et al., 1984). It is not possible, therefore, to isolate the defective enzyme complex by conventional purification procedures, all of which involve a long density gradient centrifugation or gel filtration step, following the solubilization of the ATP synthase from the mitochondrial membrane with detergents (see Tzagoloff and Meagher [1971], Ryrie and Gallagher [1979] and Rott and Nelson [1981] for example).

One of the anti-α and one of the anti-β subunits were found to be particularly effective in the immunoprecipitation of intact ATP synthase from the detergent extract in a solid-phase system (Hadikusumo et al., 1984; and Table I), and thus providing the essential quick probe of assembly for the analysis of the defective enzyme complex in the yeast mutants. As shown in Fig. 2, the subunit composition of the enzyme complex immunoprecipitated with these antibodies is essentially the same as that isolated by centrifugation in a linear glycerol gradient as described by Tzagoloff and Meagher (1971). In both cases, at least ten polypeptides can be detected to be associated with the enzyme complex. Five of these polypeptides have been shown to be the F_1 subunits α, β, γ, δ and ϵ (Hadikusumo et al., 1984). The mitochondrially synthesized subunits 6, 8 and 9 of the F_0 sector have also been identified by differential labeling of the mitochondrial translation products with [^{35}S]methionine in the presence of a cytoplasmic ribosomes inhibitor such as cycloheximide (Hadikusumo et al., 1984). In addition to the F_1 and the F_0 subunits, two other polypeptides were found to copurify with the immunoprecipitates, the apparent M_r of which in SDS-polyacrylamide gels were 25 and 18 kDa (designated P25 and P18 respectively). These polypeptides are most likely to be true subunits of the yeast ATP synthase; they were routinely co-precipitated by the monoclonal antibodies to the α and the β subunits of the enzyme complex, and with the improvement in the resolution of the SDS-polyacrylamide gel electrophoresis in recent years, could be routinely observed in the glycerol density gradient purified complex (Fig. 2c).

The cloning of the *unc* operon of E. coli has allowed detailed investigations of the molecular structure-function and the assembly of the ATP synthase in this organism. It has also become apparent in recent years, however, that the mitochondrial ATP synthases, while sharing the general features of the prokaryotic enzyme complex, are significantly more complex as reflected in their subunit composition. The ease with which yeast cells can be manipulated physiologically and genetically has made this organism an ideal model system to investigate the assembly of the mitochondrial ATP synthase. The comparison between the subunit composition of the yeast and the mammalian mitochondrial ATP synthases, therefore, is of a particular interest. It is possible that the 18 kDa subunit is indeed the oligomycin

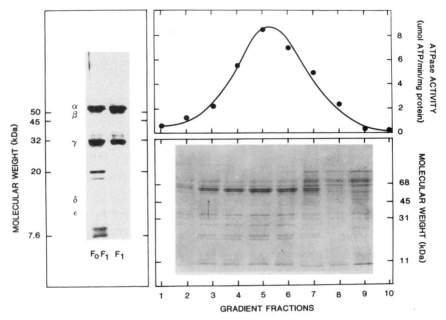

Fig. 2. Subunit Composition of the Yeast Mitochondrial ATP Synthase. F_oF_1-ATPase was solubilized from the yeast mitochondrial membrane with Triton X-100 (Tzagoloff and Meagher, 1971). The enzyme complex was then isolated from the Triton extract by either (a) immunoprecipitation with the monoclonal anti-ATP synthase β subunit RH48 (Hadikusumo et al., 1984), or (b and c) centrifugation in a glycerol gradient (Tzagoloff and Meagher, 1971). Shown are (a) the autoradiograph of [^{35}S]-labeled F_oF_1-ATPase and F_1-ATPase isolated by immunoprecipitation from the Triton extract and from a crude F_1-ATPase preparation (Beechey et al., 1975) respectively following electrophoresis in an SDS-polyacrylamide gel, (b) the ATPase specific activity of the various fractions obtained from the glycerol gradient centrifugation of the Triton solubilized F_oF_1-ATPase and (c) silver-stained pattern of each glycerol gradient fraction after electrophoresis in an SDS-polyacrylamide gel. Similar pattern to (a) was obtained when the monoclonal anti-ATP synthase α subunit was used for immunoprecipitation. The identity of the protein bands in gels is as previously reported (Hadikusumo et al., 1984).

sensitivity conferring protein (OSCP) which has been reported to be of a similar molecular weight in yeast. A polypeptide of 24 kDa has been reported to be associated with the beef heart mitochondrial ATP synthase (Walker et al., 1987). The sequence of this protein subunit (designated subunit *b*), and that of a 25 kDa protein isolated from the yeast ATP synthase, have been recently determined by direct protein sequencing and by the sequencing of cDNA clones (Walker et al, 1987; Velours et al., 1987). Examination of the sequences of the two polypeptides revealed significant homology, and their hydropathy profiles were found to resemble that of the *E. coli* subunit *b*.

In addition to the ten subunits described above, a number of other protein factors have been reported to be associated with the mammalian mitochondrial ATP synthase. These include Factor 6 (F_6), Factor B (F_B) and subunit *d* (see Walker et al., 1987). The presence of these polypeptides has not been demonstrated in yeast. It is interesting to note, however, that when the yeast ATP synthase was analyzed by SDS-polyacrylamide gel electrophoresis employing a sensitive silver staining procedure, at least one additional major band could be observed to be associated with the enzyme complex (data not shown).

THE ASSEMBLY PATHWAY OF THE YEAST MITOCHONDRIAL ATP SYNTHASE

The two monoclonal antibodies which can precipitate intact ATP synthase from the Triton X-100 extract of the yeast mitochondria have been used to define the assembly defects in *mit⁻* mutants of the yeast *Saccharomyces cerevisiae* with lesions in the mitochondrial structural genes of the ATP synthase membrane F_o-subunits 6, 8 and 9. The mutations in these strains have been defined by the sequencing of the oli1 (John et al., 1986), oli2 (Ooi et al., 1985) and aap1 (Macreadie et al., 1983) genes respectively, and the strains analyzed include mutants which do not synthesize one of the membrane subunits, as well as mutants which can synthesize these subunits but in an altered form. It was suggested from the results of these analyses, that subunit 9 plays a key role in the assembly or the stabilization of the F_o-sector, in that in the absence of this polypeptide, no subunits 6 or 8 was found to be associated with the defective ATP synthase complex (Marzuki et al., 1983; Linnane et al., 1985; Hadikusumo et al., 1988). Thus, subunit 9 is probably the first mitochondrially synthesized subunit to be assembled to the F_1-sector. The assembly of subunit 9 appears to be followed by subunit 8, because in the absence of this subunit the ATP synthase complex still contains subunit 9, but not subunit 6. Subunit 6 is apparently the last mitochondrially synthesized subunit to be assembled, since the absence of subunit 6 does not affect the assembly of subunit 8 or 9.

The previous studies (Marzuki et al., 1983; Linnane et al., 1985) were designed specifically to examine the mitochondrially synthesized subunits of the ATP synthase. The extent to which the mutations in the mit^- strains have affected the association of the cytoplasmically synthesized subunits of the enzyme complex to its β subunit has now been investigated. For this purpose, cells were labeled with [^{35}S]sulphate for four hours, and protein subunits which can still be assembled to the defective ATP synthase in the mit^- mutants were determined by immunoprecipitation using the monoclonal anti-β subunit antibody RH 48.

All the mit^- mutants studied showed the presence of the five F_1 subunits in the immunoprecipitate indicating that these subunits were assembled irrespective of the defect in the F_0 sector. The examination of the F_0 subunits associated with the immunoprecipitates confirms the previous observations (Marzuki et al., 1983; Linnane et al., 1985). The assembly of the 25 kDa and the 18 kDa subunits appears to be dependent on the proper formation of the F_0-sector. When any one of the F_0 subunits was not present, the P18 and P25 polypeptides were not observed in the ATP synthase immunoprecipitates of the mit^- mutants (see Fig. 3 for subunit 6-less mutants). Some P18 and P25 might be associated with the immunoprecipitates from mutants in which all subunits of the F_0-sector were assembled despite of their mutational defects. However, the amount of P18 and P25 polypeptides in the defective ATP synthase was significantly reduced and variable; in some experiments either one or both of these subunits could not be detected.

The revised pathway proposed for the assembly of the yeast ATP synthase, deduced from the assembly defects observed in the mit^- mutants, is shown in Fig. 4. This pathway differs significantly from that which has been proposed for the assembly of E. coli ATP synthase. The difference in the assembly sequence might be the necessary consequence of the fact that the mitochondrial ATP synthase is assembled by the coordination of two genetic systems: the mitochondrial system responsible for the synthesis of the hydrophobic F_0-subunits, and the nucleo-cytoplasmic system which synthesis the other subunits. It should be noted, however, that inherent to the many studies conducted in E. coli is the assumption that all protein subunits which are inserted into the plasma membrane are assembled into the enzyme complex.

ANTIBODIES TO THE HUMAN ATP SYNTHASE

The investigation of the biochemical and molecular defects underlying inherited disorders of the mitochondrial development has become a major interest in a number of laboratories in recent years (see Morgan-Hughes [1984], Di Mauro et al., [1985] and

Marzuki et al. [1988] for recent reviews).   Antibodies which can precipitate intact mammalian mitochondrial ATP synthase are essential to define the enzyme assembly defects in patients with mitochondrial cytopathies.   A program has been initiated, therefore, to raise subunit-specific antibodies to the human mitochondrial ATP synthase, and to other complexes of the human mitochondrial respiratory chain, with a long term objective of employing these antibodies to define enzyme assembly defects in patients with mitochondrial cytopathies. For the ATP synthase, the strategy adopted involves the synthesis of short peptides corresponding to specific regions on the β subunit of the enzyme complex, and raising antibodies to the synthetic peptides. The first peptide synthesized corresponds to the eight carboxy terminal amino acid residues of the human β subunit 7]. This region has been selected because the interaction of the antibodies raised to this part of the molecule would probably result in only a

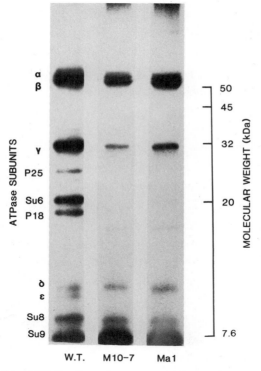

Fig. 3.   Defective ATP Synthase Immunoprecipitated from Subunit 6-less Mutants of *Saccharomyces cerevisiae* does not Contain P18 and P25. [^{35}S]-labeled ATP synthase (Hadikusumo et al., 1988) was solubilized with Triton X-100 from mitochondria of a wild-type strain and two *mit*⁻ mutants (M10-7 and Ma1) of yeast which do not synthesize subunit 6 as a result of mutations in the *oli 2* gene (Ooi et al., 1985).   The ATP synthase was then isolated by immunoprecipitation with the monoclonal anti-ATP synthase β subunit RH48, electrophoresed on a 12% SDS-polyacrylamide gel (Hadikusumo et al., 1988), and visualized by fluorography (Chamberlain, 1979). The various subunits of the enzyme complex were identified as in (Hadikusumo et al., 1984).

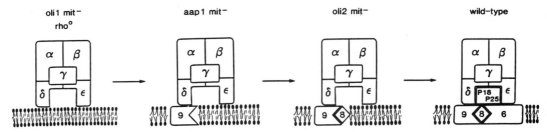

Fig. 4. Assembly Pathway of the Yeast ATP Synthase. The assembly pathway of the Fo subunits was deduced from the enzyme assembly defects observed in *oli1 mit⁻*, *aap1 mit⁻* and *oli1 mit⁻* mutants of which do not synthesize the ATP synthase subunits 9, 8 and 6 respectively. The activity of the F_1-ATPase which still can be assembled in a mtDNA-less *rho⁰* strain was found to be modulated by the fatty acid composition of the mitochondrial membrane (Orian et al., 1984), and thus is presumably already associated with the mitochondrial inner membrane. Reproduced from Hadikusumo et al., 1988).

minimal effect on the conformation of the β subunit; this is an important consideration as most of the monoclonal antibodies which have been raised against the yeast β subunit induce the dissociation of some of the F_1-subunits (Hadikusumo et al., 1984).

The property of rabbit antibodies to the synthetic peptide is shown in Fig. 5. The rabbit sera contained significant levels of antibodies to the synthetic peptide, as well as antibodies to the immunization carrier, keyhole limpet hemocyanine, as expected; when tested by ELISA, the antisera reacted with a peptide-bovine serum albumin conjugate (A_{405} = 0.6 at a dilution of 1:1000), but not with bovine serum albumin only (Fig. 5b). The antibodies showed a good reactivity to the human F_1-ATPase in ELISA, presumably to the β subunit. Examination of the rabbit sera by Western immunoblotting against a human F_1-ATPase preparation indeed revealed only one band (Fig. 5a) with the mobility of the β subunit (56 kDa). The antibodies showed some reactivity to the bovine and the yeast F_1-ATPase in the ELISA, but only weakly (A_{405} around 0.35 and 0.15 for bovine and yeast respectively, compared to 1.1 for human F_1-ATPase at a dilution of 1:100; Fig. 5b). Similarly, only weak reactivity could be detected when the antibodies were examined by Western immunoblotting against the bovine and yeast F_1-ATPase (Fig. 5a). This observation was expected for the yeast β subunit, considering that the yeast subunit shared only four common amino acid residues with the human subunit within the stretch of eight residues in the carboxy terminal (see legend to Fig. 5). It is of interest to note, however, that the antibodies recognize only very weakly the bovine subunit which differ from the human polypeptide by only one additional serine at the carboxy terminal.

The ability of the antipeptide antibodies to immunoprecipitate intact human ATP synthase has been examined in a solid-phase system. For this purpose, the antibodies were first conjugated to cyanogen bromide-activated Sepharose 4B as described for the yeast monoclonal antibodies to the ATP synthase α and β subunits (Hadikusumo et al., 1984). The human ATP synthase was solubilized from the mitochondrial membrane with octyl-glucoside (Rott and Nelson, 1981). Preliminary results of this investigation indicated that the anti-human ATP synthase β subunit can immunoprecipitate the intact complex. The subunit composition of the precipitated human ATP synthase was found to be very similar to that of the yeast enzyme. Beside the established subunits of the F_1- and the F_0-sectors, at least two additional bands were observed with the apparent mobility of a 29 kDa and an 18 kDa proteins, presumably the equivalent of the yeast P25 and P18.

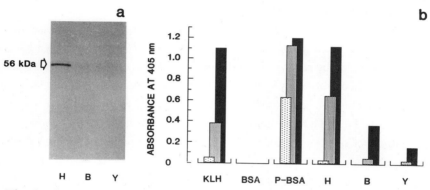

Fig. 5. Antipeptide Antibodies to the Carboxy Terminal Region of the Human ATP Synthase. A peptide of eight amino acid residues, corresponding to the carboxy terminal of the human ATP synthase β subunit (Ohta and Kagawa, 1986) was synthesized and conjugated to keyhole limpet hemocyanine for immunization in rabbits and mice. Shown are the properties of the antibodies from a rabbit, characterized by (a) Western–immunoblotting with F_1-ATPase preparations and (b) Enzyme Linked ImmunoSorbent Assay. The cross-reactivity of the antibodies with bovine and yeast ATP synthase was also determined. The carboxy terminal sequence of human, bovine and yeast ATP synthase β subunits is as follows (the synthetic peptide sequence is underlined):

Human - A V A K A D <u>K L A E E H S S</u>-COOH (Ohta and Kagawa, 1986)

Bovine - A V A K A D K L A E E H S-COOH (Walker et al., 1985)

Yeast - V V R K A E K L A R E A N-COOH (Takeda et al., 1985)

H = human F_1-ATPase, B = bovine F_1-ATPase, Y = yeast F_1-ATPase, KLH = keyhole limpet hemocyanine, BSA = bovine serum albumin, and P-BSA = the synthetic peptide conjugated to bovine serum albumin. The molecular weights indicated were determined from the electrophoretic mobilities of standard proteins run on a parallel track.

CONCLUDING REMARKS

The usefulness of epitope-specific antibodies in the study of the structure, function and assembly of the ATP synthase is well illustrated in the present communication. In particular, monoclonal antibodies to subunits of the yeast ATP synthase have been instrumental in the definition of the subunit composition of the enzyme complex, and in the elucidation of its assembly pathway. Many other important questions can be addressed using these antibodies. The molecular definition of the epitopes recognized by the monoclonal antibodies, for example, would lead to important information with regard to the catalytic regions on the ATP synthase. Molecular mapping of the various subunits assembled in the ATP synthase by electron microscopy (Lunsdorf et al., 1984) with antibodies with defined epitopes would also give significant information on the tertiary folding of the subunits, and their organization in the F_oF_1-ATPase.

An investigation into the assembly defects of the human ATP synthase in various mitochondrial disorders has been initiated with a similar approach to that employed for the yeast mutants. For this purpose, anti-human ATP synthase β subunit has been produced by raising antibodies to synthetic peptides. ATP synthase assembly defects might occur not only in patients with mutations in one of the structural genes for enzyme subunits, but could be also secondary to defects in other respiratory enzyme complexes. The anti-human ATP synthase antibodies will also be useful in the investigation of other human disorders in which the involvement of ATP synthase has been implicated. For example, the autoantibodies found characteristically in sera of patients with primary biliary cirrhosis, an autoimmune disease affecting primarily the intrahepatic biliary tract, have been suggested to be directed against protein subunits of the ATP synthase (Leoutsakos et al., 1982; Lindenborn-Fotinos et al., 1982). The use of the anti-human ATP synthase β subunit to immunopurify the enzyme complex from human liver and heart mitochondria has allowed an unequivocal demonstration that the autoantigens of primary biliary cirrhosis are in fact not associated with the ATP synthase (unpublished observation). The definition of the ATP synthase assembly defects in human mitochondrial disorders is not only of significance for our understanding of the pathogenesis of these disorders, but might also lead to important information with regard to the enzyme assembly process.

REFERENCES

Beechey, R.B., Hubbard, S.A., Linnett, P.E., Mitchell, A.D. and Munn, E.A. (1975) Biochem. J. 148, 533-537.

Byrne, E., Marzuki, S. and Dennett, X. (1988) Med. J. Aust. 149, 30-33.

Chamberlain, J.P. (1979) Anal. Biochem. 98, 132-135.

Di Mauro, S., Bonilla, E., Zeviani, M., Nakagawa, M. and De Vivo, D.C. (1985) Ann. Neurol. 17, 521-538.

Dunn, W.D., Tozer, R.G., Antzak, D.F. and Heppel, L.A. (1985) J. Biol. Chem. 260, 10418-10425.

Hadikusumo, R., Hertzog, P.J. and Marzuki S. (1986) Biochim. Biophys. Acta 850, 33-40.

Hadikusumo, R.G., Hertzog, P.J. and Marzuki, S. (1984) Biochim. Biophys. Acta 765, 258-267.

Hadikusumo, R.G., Meltzer, S., Choo, W.M., Jean-Francois, M.J.B., Linnane, A.W. and Marzuki, S. (1988) Biochim. Biophys. Acta, 933, 212-222.

Jean-Francois, M.J.B., Hertzog, P.J. and Marzuki, S. (1988) Biochim. Biophys. Acta, 223-228.

John, U.P., Willson, T.W., Linnane, A.W. and Nagley, P. (1986) Nucl. Acids Res. 14, 7437-7451.

Joshi, S., Kantham, L., Kaplay, W. and Sanadi, D.R. (1985) FEBS Lett. 179, 143-147.

Leoutsakos, A., Palmer, C. and Baum, H. (1982) Prog. Clin. Biol. Res. 102, 491-503.

Lindenborn-Fotinos, J., Sayers, T.J. and Berg, P.A. (1982) Clin. Exp. Immunol. 50, 267-274.

Linnane, A.W., Lukins, H.B., Nagley, P., Marzuki, S., Hadikusumo, R.G., Jean-Francois, M.J.B., John, U.P., Ooi, B.G., Watkins, L., Willson, T.A., Wright, J. and Meltzer, S. (1985) in Achievements and Perspectives in Mitochondrial Research: 1. Bioenergetics (Quagliariello, E., Slater, E.C., Palmieri, F., Saccone, C. and Kroon, A.M. eds.) pp. 211-221, Elsevier, Amsterdam.

Lunsdorf, H., Ehrig, K., Friedl, P. and Schrairer, H.V. (1984) J. Mol. Biol. 173, 131-136.

Macreadie, I.G., Novitski, C.E., Maxwell, R.J., John, U., Ooi, B.G., McMullen, G.L., Lukins, H.B., Linnane, A.W. and Nagley, P. (1983) Nucl. Acids Res. 11, 4435-4451.

Marzuki S., Hadikusumo, R.G., Choo, W.M., Watkins, L., Lukins, H.B. and Linnane, A.W. (1983) in Mitochondria 1983. Nuclear Cytoplasmic Interactions (Wolf, K., Schweyen, R.Y. and Kaudewitz, F. eds.) pp. 535-549, De Gruyter, Berlin.

Marzuki, S. and Linnane, A.W. (1982) Methods Enzymol. 97, 294-305.

Moradi-Ameli, M. and Godinot, C. (1983) Proc. Natl. Acad. Sci. USA 80, 6167-6171.

Morgan-Hughes, J.A., Hayes, D.J. and Clark, J.B. (1984) in Neuromuscular Diseases (Serratrice, G. et al. Eds.) pp. 79-85, Raven Press, New York.

O'Farrel, P.H. (1975) J. Biol. Chem. 250, 4007-4021.

Ohta, S. and Kagawa, Y. (1986) J. Biochem. 99, 135-141.

Ooi, B.G., McMullen, G.L., Linnane, A.W., Nagley, P. and Novitski, C.E. (1985) Nucl. Acids Res. 13, 1327-1339.

Orian, J.M., Hadikusumo, R., Marzuki, S., and Linnane, A.W. (1984) J. Bioenerg. Biomemb. 16, 561-581.

Rott, R. and Nelson, N. (1981) J. Biol. Chem. 256, 9224-9228.

Ryrie, I.J. and Gallagher, A. (1979) Biochim. Biophys. Acta 545, 1-14.

Sanadi, D.R. (1982) Biochim. Biophys. Acta 683, 39-56.

Takeda, M., Vassarotti, A. and Douglas, M. (1985) J. Biol. Chem. 260, 15458-15465.

Tzagoloff, A. and Meagher, P. (1971) J. Biol. Chem. 246, 7328-7336.

Velours, J., Arselin-de Chateaubodeau, G., Galante, M. and Guerin, B. (1987) Eur. J. Biochem. 164, 579-584.

Walker, J.E., Fearnley, I.M., Gay, N.J., Gibson, B.W., Northrop, F.D., Powell, S.J., Runswick, M.J., Saraste, M. and Tybulewicz, V.L.J. (1985) J. Mol. Biol. 184, 677-701.

Walker, J.E., Runswick, M.J. and Poulter, L. (1987) J. Mol. Biol. 197, 89-100.

THE PROTON-ATPase OF CHROMAFFIN GRANULES AND

SYNAPTIC VESICLES

Yoshinori Moriyama and Nathan Nelson*

Roche Institute of Molecular Biology, Roche Research Center
Nutley, New Jersey 07110

Proton-ATPases can be divided into three main classes: (1) Plasma membrane-type, which operates via a phosphoenzyme intermediate, and therefore is part of the E_1-E_2 ion pumps. (2) The mitochondrial-type, which is present in eubacteria, mitochondria and chloroplasts. It operates without phosphoenzyme intermediate and its main function is to phosphorylate ADP at the expense of proton motive force. (3) The vacuolar-type ATPase, which is present in the vacuolar system of eukaryotic cells and probably in archaebacteria. It functions in a controlled pumping of protons into organelles of the vacuolar system via unknown mechanism that presumably does not involve phosphoenzyme intermediate.

The vacuolar proton-ATPase is a landmark for the secretory pathway of eukaryotic cells. It was detected in several organelles connected with this system; for example, lysosomes (Schneider, 1981; Moriyama et al., 1984; Reeves, 1984; Schneider, 1987), synaptic vesicles (Cidon et al., 1983; Toll and Howard, 1980), chromaffin granules (Cidon and Nelson, 1983; Cidon and Nelson, 1986), clathrin-coated vesicles (Xie and Stone, 1986; Forgac et al , 1983; Stone et al., 1983), plant and fungal vacuoles (Uchida et al, 1985; Bowman et al., 1986; Mandala and Taiz, 1985; Manolson et al., 1985; Randall and Sze, 1986; Marin et al., 1985), platelet dense granules (Dean et al., 1984; Rudnick, 1986) and the Golgi complex (Chanson and Taiz, 1985; Ali and Akazawa, 1986). We purified the proton-ATPase from chromaffin granule membranes and studied its enzymatic properties following reconstitution into lipid vesicles.

THE PROTON ATPase FROM CHROMAFFIN GRANULES

The ATPase activity of chromaffin granule membranes was discovered over two decades ago (Kirshner, 1962). Since then a large body of evidence has clearly shown that this ATPase is a proton pump responsible for generating proton motive force for catecholamine uptake (Pollard et al., 1976; Casey et al., 1977; Flatmark and Ingebretsen, 1977; Phillips and Allison, 1978; Schuldiner et al., 1978; Johnson et al., 1978; Johnson and Scarpa, 1979; Kanner et al., 1980). The identity of the enzyme was not clear, and most of the purification attempts yielded preparations containing an enzyme similar to the mitochondrial proton-ATPase (Apps and Glover, 1978; Apps and Schatz, 1979; Roisin and Henry, 1982). Using immunological and other techniques, it was demonstrated that the mitochondrial proton-ATPase was present in the chromaffin granule preparations due to contamination of mitochondrial membranes (Cidon and Nelson, 1983; Cidon and Nelson, 1982; Cidon et al., 1983). A novel proton-ATPase, that was not homologous to the mitochondrial enzyme, was identified in the chromaffin granule membranes. Recently, the enzyme was purified and it was shown that previous attempts to purify the enzyme were hampered by its instability that caused loss of activity following detergent treatment (Cidon and Nelson, 1986). By including the protease inhibitors pepstatin A and leupeptin and decreasing the time required for the purification of the enzyme to about 10 h, the enzyme was purified in a state which allowed reconstitution into vesicles having ATP-dependent proton uptake activity (Moriyama and Nelson, 1987). Fig. 1 shows SDS polyacrylamide gel of the purified proton-ATPase complex from chromaffin granule membranes. The purified preparation contains five major protein bands denoted as subunits I to V in the order of decreasing molecular weights of 115 kDa, 72 kDa, 57 kDa, 39 kDa and 17 kDa, respectively. Polypeptides at positions of 41, 34, 33 and 20 kDa are also detected. Further studies are required to determine the integrity of each polypeptide in the protein complex, and the function of each individual subunit in the various activities of the enzyme.

The ATP-dependent proton uptake activity of chromaffin granule membranes is sensitive to NEM-treatment. However, the ATPase activity of the same preparation was reported to have much lower affinity to the inhibitor (Flatmark et al., 1982). The same phenomenon was observed with the purified enzyme following reconstitution into lipid vesicles (not shown). Labeling experiments with ^{14}C-NEM revealed that subunits I, II and IV are labeled with the alkylating agent (Fig. 2). Increasing amounts of cold NEM diminished the labeling of subunit II (72 kDa), while subunits I (115 kDa) and IV (39 kDa) were still labeled (Moriyama and Nelson, 1987). From the data depicted in Fig. 2, it is apparent that the site with the highest affinity to NEM is situated on the 72 kDa polypeptide and this site is exposed to the aqueous environment. Pretreatment with cold DIUS specifically prevented the labeling of this subunit. Initially, it appears that the 72 kDa subunit functions in the proton translocation activity of the enzyme.

Unlike the other two classes of ATPases, the vacuolar-type enzymes are influenced greatly by the presence of anions. We demonstrated that the proton uptake activity of the reconstituted proton-ATPase from chromaffin granules was absolutely dependent on the presence of Cl^- or Br^- in the medium (Moriyama and Nelson, 1987). Sulfate inhibits the proton uptake by competing with chloride at a site located on the outer face of the membrane. The effect of nitrate on proton uptake activity is rather complicated. As with the inhibition by NEM, proton pumping activity was much more sensitive to nitrate than the ATPase activity. About 20 mM nitrate was sufficient for 90% inhibition of the proton pumping activity, while at 100 mM only 50% of the ATPase activity was inhibited both *in situ* and following reconstitution of the purified enzyme.

Fig. 1. SDS Gel of the Purified H^+-ATPase from Chromaffin Granule Membranes. H^+-ATPase was purified according to Moriyama and Nelson (1987) and electrophoresed on 10% polyacrylamide gel containing SDS. The gel was stained with Coomassie brilliant blue. Five major bands were designated as subunit I to V according to the order of the molecular weights of 115, 72, 57, 39 and 17 kDa, respectively.

Inclusion of nitrate during reconstitution of the enzyme abolished the proton uptake activity of the system. As shown in Fig. 3, addition of nitrate to reconstituted vesicles first stimulated the proton uptake and then inhibited the activity. This was not due to changes in membrane potential, which was collapsed by the inclusion of valinomycin and potassium. As illustrated in Fig. 4, this inhibition is reversible and preloading of the vesicles with nitrate, while assaying them in the presence of external chloride causes a lag in the proton uptake activity, which decreased with time. These effects suggest the presence of anion binding sites on the enzyme, both outside and

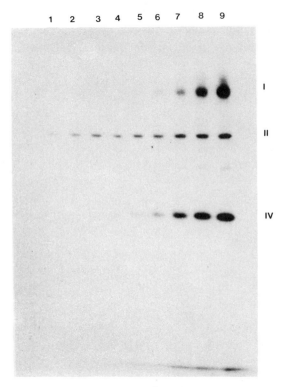

Fig. 2. Labeling of the Reconstituted H^+-ATPase by $[^{14}C]$-N-ethylmaleimide. The enzyme was purified and reconstituted into liposomes according to Moriyama and Nelson (1987). Thioglycerol and ATP were omitted from the last steps of purification and the reconstitution. Proteoliposomes containing active H^+-ATPase were incubated for 10 min at room temperature in 50 μl solution consisting of 20 mM MOPS-Tris pH 7.0, 0.1 M KCl and various amounts of ^{14}C-NEM. In all of the treatment the specific activity of ^{14}C-NEM was kept at 2 μCi/μmole. The NEM concentrations were the following: (1) 5 μM, (2) 10 μM, (3) 15 μM, (4) 20 μM, (5) 30 μM, (6) 60 μM, (7) 75 μM, (8) 100 μM and (9) 150 μM. The reaction was terminated by addition of 10 μl dissociating solution containing 10% SDS and 10% β-mercaptoethanol. Thirty-five μl of the samples were applied on 10% SDS gel and electrophoresed. The gel was fluorographed by Amplify (Amersham).

inside the vesicles. The proton uptake activity is dependent on binding of chloride on the outer site, while nitrate inhibits exclusively when bound to the internal site.

Accumulation of protons in the reconstituted vesicles was dependent on the presence of potassium and valinomycin during reconstitution. The presence of chloride could not alleviate this dependency. In isolated membranes, chloride by itself is sufficient to facilitate the ATP-dependent activity of the system (Cidon et al., 1983). The chloride channel was completely eliminated during the purification of the enzyme. Apparently, proton uptake activity is very sensitive to positive membrane potential inside the vesicles, and proton pumping activity is inhibited before thermodynamic equilibrium is reached. It is likely that membrane potential, together with anions, are the main factors involved in regulation of the pH gradient across the membranes of the various organelles containing this kind of ATPase. This kinetic control prevents over acidification of the organelles. The amounts of various anions in different cell types can be modulated by specific anion transports and receptors like the muscarinic acetylcholine receptor. Recently we observed that the proton-ATPase of chromaffin granules is an allosteric enzyme (Moriyama and Nelson, unpublished), and like the eubacterial type enzyme may possess more than one nucleotide binding site (Nelson, 1976). Fig. 5 shows that following UV illumination, $\alpha[^{35}S]$-ATP and 8-azido ATP

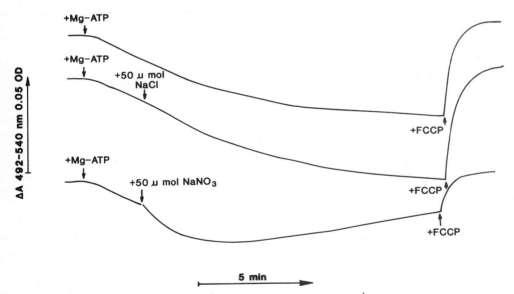

Fig. 3. Effect of NaCl and NaNO$_3$ on the Reconstituted H$^+$-Pumping Activity. Purified enzyme (5 μl, 1.5 μg protein) was reconstituted by dilution in 1 ml of solution containing 20 mM MOPS-tris pH 7.0, 20 mM KCl and 0.1 μg valinomycin. Then 15 μl of 1 mM acridine orange were added. ATP dependent H$^+$-pump activity was measured by its absorbance changes at 492–540 nm. The H$^+$-pump activity was initiated by addition of MgATP (1 mM) and terminated by FCCP (1 μM).

labeled three subunits of the enzyme of 115, 72 and 39 kDa. Competition experiments revealed that the 72 kDa polypeptide contains the nucleotide binding site with the highest affinity. Further studies are required to determine whether the labeling of three subunits is due to the presence of three different binding sites or if it is due to the long half life of the radical formed by the UV illumination. Fig. 6 depicts a model for the subunit structure and function of the vacuolar proton-ATPases. Subunits A to E, which comprise the catalytic sector of the enzyme, are polypeptides of 72, 57, 41, 34 and 33 kDa. The 20 and 16 kDa subunits are part of the membrane sector. The polypeptides of 115 and 39 kDa found in vacuolar ATPases of mammalian sources are accessory subunits.

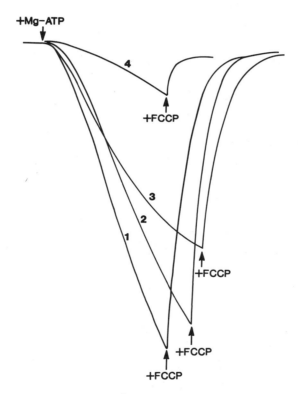

Fig. 4. Reversible Inhibition of H^+-pump Activity by Nitrate. Purified enzyme (100 μl, 33 μg protein) was reconstituted into 20 ml of solution containing 20 ml MOPS-tris pH 7.0, 1 mM dithiothreitol and KCl 0.1 M (1.3) or KNO_3 0.1 M (traces 2 and 4). Proteoliposomes were collected by ultracentrifugation at 200,000 g for 1 hr and suspended in 20 mM MOPS-tris, 0.1 M KCl and 1 mM DTT. ATP dependent H^+-pump activity was measured as described in the legend of Fig.3, except that in 3 and 4, KNO_3 was substituted for KCl in the assay solution.

PROTON ATPase OF SYNAPTIC VESICLES

It has been known for some time that synaptic vesicles acidify their interior through the action of a proton-ATPase (Toll and Howard, 1980). In previous studies we had analyzed ATP-dependent proton uptake in crude synaptosome preparations (Cidon et al., 1983). Synaptic vesicles from rat brain were purified (Huttner et al., 1983) and proton uptake activity of the preparation was studied. The specific activity was about 5-fold greater than in the synaptosome preparation, and it is quite likely that the activity in synaptosomes was due to the presence of synaptic vesicles in the preparation. As yet, the enzyme has not been purified but its properties suggest it resembles the enzyme from chromaffin granules (Moriyama and Nelson, unpublished).

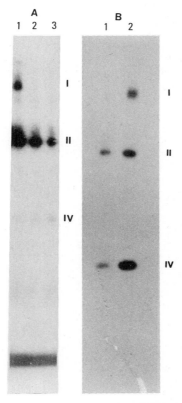

Fig. 5. Photoaffinity Labeling of Chromaffin H^+-ATPase. A. Labeled with [α-2P]8-azido ATP. Purified enzyme (3.5 μg) was labeled with [α-^{32}P]8-azido ATP (5.2 Ci/mmol, ICN) under UV illumination for 10 min on ice. (1) Solution consisting of 10 mM MOPS-tris pH 7.0, 50 mM KCl, 5.6 μM 8-azido ATP and enzyme plus $MgCl_2$ 2 mM. (2) As in 1 plus ATP 1 mM. (3) As in 1 plus $MgCl_2$ 2 mM and ATP 1 mM. B. Labeled with [α-^{35}S]ATP. Reconstituted enzyme (2.4 μg) was labeled with [α-^{35}S]ATP (650 Ci/mmol, Amersham) under UV illumination for 10 min on ice. (1) Solution containing 10 mM MOPS-tris pH 7.0, 50 mM KCl, 0.15 μM [α-^{35}S]ATP and enzyme. (2) As in 1 plus $MgCl_2$ 2 mM.

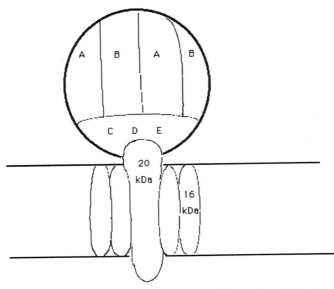

Fig. 6. A model for the Subunit Structure and Function of the Chromaffin Granule H^+-ATPase.

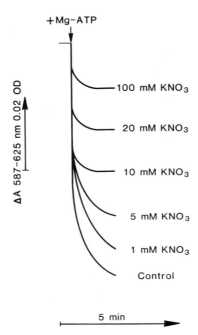

Fig. 7. Effect of Nitrate on the Membrane Potential Formed by H^+-ATPase in Synaptic Vesicles. Membrane potential was measured by absorption change of oxonol VI at 587-625 nm. Assay mixture contained 20 mM MOPS-tris (pH 7.0), 0.2 M sucrose, synaptic vesicle containing 120 μg protein, and KNO_3 as specified. Mg-ATP (1 mM) was added to initiate ATP dependent H^+-pump activity.

Like in chromaffin granules, the ATP-dependent proton uptake activity of synaptic vesicles is sensitive to nitrate. As shown in Fig. 7, even the formation of ATP-dependent membrane potential was abolished by nitrate. This effect may be the result of both inhibition of the enzyme activity and collapse of the membrane potential by influx of nitrate through the chloride channel. Fig. 8 compares the effect of chloride

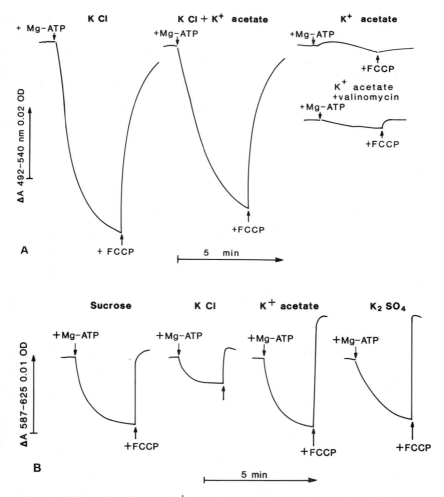

Fig. 8. Effect of Anions on H^+-pumping Activity and Formation of Membrane Potential in Synaptic Vesicles. Top: Proton uptake. Proton uptake was measured by acridine orange absorbance change at 492–540 nm. Assay solution consisted of 20 mM MOPS-tris pH 7.0, 0.1 M of the listed salts, and synaptic vesicles containing 40 μg protein. Where indicated, Mg-ATP 1 mM, FCCP 1 μM, valinomycin 0.1 μg were added. Bottom: Membrane potential. Membrane potential was measured by oxonol VI absorbance change at 587–625 nm. Assay solution was 20 mM MOPS-tris pH 7.0, 0.2 M sucrose, 5 μM oxonol VI and 40 μg protein. Listed salts were also added as 0.1 M solution. H^+-pump activity was initiated by the addition of Mg-ATP (1 mM) and terminated by FCCP (1 μM). A, proton uptake. B, membrane potential.

and acetate on ATP-dependent proton uptake activity and formation of membrane potential. Chloride was required for ATP-dependent proton uptake of the preparation. The presence of potassium and valinomycin was not necessary although they enhanced the activity (not shown). Acetate could not replace chloride and it inhibited the proton uptake activity in the presence of chloride. Membrane potential was formed even without the presence of anions and as expected chloride decreased the extent of membrane potential while inducing proton uptake. Recently we raised an antibody against the 115 kDa subunit of the proton-ATPase from chromaffin granules. This antibody crossreacted with a polypeptide of the same molecular weight in synaptic vesicles. Thus, the presence of a proton-ATPase similar to the one purified from chromaffin granules in synaptic vesicles is very likely.

REFERENCES

Ali, M.S. and Akazawa, T. (1986) Plant Physiol. 81, 222-227.

Apps, D.K. and Glover, L.A. (1978) FEBS Lett. 85, 254-258.

Apps, D.K. and Schatz, G. (1979) Eur. J. Biochem. 100, 411-419.

Bowman, E.J., Mandala, S., Taiz, L. and Bowman, B.J. (1986) Proc. Natl. Acad. Sci. USA 83, 48-52.

Casey, R.P., Njus, D., Radda, G.K. and Sehr, P.A. (1977) Biochemistry 16, 972-977.

Chanson, A. and Taiz, L. (1985) Plant Physiol. 78, 232-240.

Cidon, S. and Nelson, N. (1982) J. Bioenerg. Biomembr. 14, 499-512.

Cidon, S. and Nelson, N. (1983) J. Biol. Chem. 258, 2892-2898.

Cidon, S. and Nelson, N. (1986) J. Biol. Chem. 261, 9222-9227.

Cidon, S., Ben-David, H. and Nelson, N. (1983) J. Biol. Chem. 258, 11684-11688.

Dean, G.E., Fishkes, H., Nelson, P.J. and Rudnick, G. (1984) J. Biol. Chem. 259, 9569-9574.

Flatmark, T. and Ingebretsen, O.C. (1977) FEBS Lett. 78, 53-56.

Flatmark, T., Gronberg, M., Husebye, E. and Berge, S.V. (1982) FEBS Lett. 149, 71-74.

Forgac, M., Cantley, L., Wiedenmann, B., Altstil, L. and Branton, D. (1983) Proc. Natl. Acad. Sci. USA 80, 1300-1303.

Huttner, W.B., Schiebler, W., Greengard, P. and DeCamilli, P. (1983) J. Cell Biol. 96, 1374-1388.

Johnson, R.G. and Scarpa, A. (1979). J. Biol. Chem. 254, 3750-3760.

Johnson, R.G., Carlson, N.J. and Scarpa, A. (1979) J. Biol. Chem. 253, 1512-1521.

Kanner, B.I., Sharon, I., Maron, R. and Schuldiner, S. (1980) FEBS Lett. 111, 83-86.

Kirshner, N. (1962) J. Biol. Chem. 237, 2311-2317.

Mandala, S. and Taiz, L. (1985) Plant Physiol. 78, 327-333.

Manolson, M.F., Rea, P.A. and Poole, R.J. J. (1985) J. Biol. Chem. 260, 12273-12279.

Manolson, M.F., Rea, P.A. and Poole, R.J. J. (1985) J. Biol. Chem. 260, 12273-12279.

Marin, B., Preisser, J. and Komor, E. (1985) Eur. J. Biochem. 151, 131-140.

Moriyama, Y. and Nelson, N. (1987) J. Biol. Chem., in press.

Moriyama, Y., Takano, T. and Ohkuma, S. (1984) J. Biochem. (Tokyo) 95, 995-1007.

Nelson, N. (1976) Biochim. Biophys. Acta 456, 314-338.

Phillips, J.H. and Allison, Y.P. (1978) Biochem. J. 170, 661-672.

Pollard, H.B., Zinder, O., Hoffman, P.G. and Nikodejevic, O. (1976) J. Biol. Chem. 251, 4544-4550.

Randall, S.K. and Sze, H. (1986) J. Biol. Chem. 261, 1364-1371.

Reeves, J.P. (1984) in Lysosomes in Biology and Pathology (Dingle, J.T., Dean, R.T. and Sly, W., Eds.) pp. 175-199, Elsevier, Amsterdam.

Roisin, M.P. and Henry, J.P. (1982) Biochim. Biophys. Acta 681, 292-299.

Rudnick, G. (1986) Ann. Rev. Physiol. 48, 403-413.

Schneider, D.L. (1987) Biochim. Biophys. Acta, in press.

Schneider, D.L. (1981) J. Biol. Chem. 256, 3858-3864.

Schuldiner, S., Fishkes, H. and Kanner, B.I. (1978) Proc. Natl. Acad. Sci. USA 75, 3713-3716.

Stone, D.K., Xie, X.-S. and Racker, E. (1983) J. Biol. Chem. 258, 4059-4062.

Toll, L. and Howard, B.D. (1980) J. Biol. Chem. 255, 1787-1789.

Uchida, E., Ohsumi, Y. and Anraku, Y. (1985) J. Biol. Chem. 260, 1090-1095.

Xie, X.-S. and Stone, D.K. (1986) J. Biol. Chem. 261, 2492-2495.

A H⁺-TRANSLOCATING ATP SYNTHASE IN AN EXTREMELY HALOPHILIC

ARCHAEBACTERIUM*

Yasuo Mukohata, Manabu Yoshida, Masaharu Isoyama,
Yasuo Sugiyama, Ayumi Fuke, Hisashi Hashimoto, Toshihiko Nanba
and Kunio Ihara

Department of Biology, Faculty of Science, Osaka University
Toyonaka 560, Japan

ABSTRACT

The H^+-translocating ATP synthase functioning on the plasma membrane of *Halobacterium halobium* is found to be different from F_oF_1-ATPase/synthase which has been believed to be ubiquitous as the proton-motive ATP synthase in all respiring organisms on our biosphere.

The synthase includes the 320 kDa component possibly as the catalytic part (ATPase when released from the membrane) which is composed of two pairs of 86 kDa and 64 kDa subunits, and the 78 kDa (and/or 12 kDa) subunit possibly as the membrane/anchor part.

INTRODUCTION

Extremely halophilic archaebacteria, halobacteria, live in nearly saturated saline by metabolizing amino acids or carbohydrates. Potassium ion is accumulated as an osmoregulatory cation in the cells and the overall intracellular salt concentration is said to be saturated.

Therefore, those machineries functioning in energy metabolism of halobacteria, esp. respiratory chain components, ATP synthase, light-energy transducing retinal

---

*This work was supported partly by Grant-in-Aid for Scientific Research on Priority Areas of "Bioenergetics" to Y.M. from the Ministry of Education, Science and Culture, Japan.

proteins and sym- and anti-ports, are of great interest not only as archaebacterial machineries of energy transduction but also as enzyme proteins functioning in such circumstances.

We initiated the present work by the finding (Matsuno-Yagi and Mukohata, 1977) and naming (Mukohata et al., 1980) of halorhodopsin, and attempted to elucidate the coupling mechanism between halorhodopsin and ATP synthesis, and at the same time to explore these individual energy transducers. Now, we found the distinct difference between the ATP synthesis of halobacteria and that carried out by F_oF_1-ATPase/synthase. In this article, we summarized the results obtained in these years on the unique and novel ATP synthase in halobacteria.

MATERIALS AND METHODS

Halobacterium halobium R_1mR (bR^-, hR^+) was cultured as described (Matsuno-Yagi and Mukohata, 1977), and cell envelope vesicles were prepared by sonication as reported (Mukohata et al., 1986). At the final sonication stage, the vesicles were stuffed with the reaction mixture at the optimum conditions for ATP synthesis; 1 M NaCl, 80 mM $MgCl_2$, 10 mM PIPES pH 6.8, 3 mM ATP and 12 mM Pi (Mukohata and Yoshida, 1987a).

The air-tight reaction vessel was furnished with a jacket for circulation of water at 30°C. Illumination was provided by a 750 W slide projector (10^5 lux) through a glass filter (> 500 nm).

The pH of the reaction medium (for intact cells, the nutrient-free culture medium and for vesicles the substrate-free reaction mixture was used) was monitored by a combination electrode and the membrane potential by the tetraphenyl-phosphonium (TPP) electrode method (Kamo et al., 1979). The ΔpH value was shown by the nominal value; the difference between the pH of the reaction medium and that of the reaction mixture stuffed in the vesicles. The membrane potential (Δψ) was not corrected for the non-specific adsorption of TPP.

The amounts of ATP in intact cells were determined under aerobic/anaerobic and/or light/dark conditions by a luciferin-luciferase method as reported (Matsuno-Yagi and Mukohata, 1977). The amounts of ATP in the stuffed vesicles were determined after ATP was synthesized by illumination and/or by the base-acid transition (the outside acidic pH jump; ΔpH = 2.8; Mukohata et al., 1986).

The membrane-bound ATPase was isolated from the envelope vesicles as reported (Nanba and Mukohata, 1987). The ATPase activity was measured at 38°C under the optimum conditions; 1.5 M Na_2SO_4, 10 mM $MnCl_2$, 40 mM MES at pH 5.8 and 4 mM ATP. The liberated Pi was determined by the established method (Taussky and Shorr, 1953). The ATPase activity of the vesicles was assayed by the above procedures after the vesicle components were solubilized with a nonionic detergent, nonaethyleneglycoldodecyl-ether, $C_{12}E_9$ (0.025%/mg protein).

ATP and ADP were purchased from Yamasa Shoyu Co., NBD-Cl and NEM from Sigma, a luciferin-luciferase ATP assay kit from Boeringer and other chemicals of reagent grade from Nakarai Chemicals except for NaCl in culture media from Japan Tobacco Industry and Tap water in culture media.

RESULTS AND DISCUSSION

Energetics of Halobacteria

Halobacterium halobium is a strictly aerobic archaebacterium which grows on peptone in culture media with vigorous aeration. The central compound in its energetics is ATP. The intracellular ATP content decreased when the air over the cell suspension in the reaction vessel was purged by nitrogen (Fig. 1). However, even under this anaerobic conditions, the ATP level was restored when the cell suspension was illuminated. This is because the machinery of ATP synthesis was driven by the proton motive force (pmf) which was built up by light-energized halorhodopsin (and/or bacteriorhodopsin; for R_1mR, not in the case). The up-down shift of the ATP content could be repeated by aerobic/anaerobic and/or light/dark switching, unless DCCD irreversibly inhibited the ATP synthesis (Mukohata and Kaji, 1981a). The DCCD binding polypeptides were identified to have the M_r sizes of 78 kDa and 12 kDa (Mukohata et al., 1987).

ATP Synthesis in Cells and Vesicles

The cell envelope vesicles of *Halobacterium halobium* R_1mR (bR^-, hR^+) were shown to synthesize ATP by the $\Delta\psi$ (Lindley and MacDonald, 1979) which was formed by illumination of halorhodopsin chloride pump (Schobert and Lanyi, 1982), and by the ΔpH which was formed by the base-acid transition (Mukohata et al., 1986). The threshold values of $\Delta\psi$ and ΔpH were -100 mV (inside negative) and 1.6 pH unit, respectively (Fig. 2). The additivity of these two driving forces was also shown qualitatively. Therefore, the machinery of the ATP synthesis in halobacteria is likely to be a H$^+$-translocating ATPase obeying the chemiosmotic energetics (Mukohata et al., 1986).

DCCD also inhibited the vesicular ATP synthesis with the half-max inhibition at 25 μM (Mukohata et al., 1986), while azide (the specific inhibitor for F_oF_1-ATPase/synthase at 5 mM) or vanadate (for E_1E_2-ATPase at 5 mM) did not (Mukohata and Yoshida, 1987a). The ATP synthesis was optimum at pH 6.8 and required Mg^{2+} more than 50 mM. The apparent Km value for ADP was around 300 μM (Mukohata and Yoshida, 1987a).

The Membrane-Bound ATPase

An ATPase was isolated from the inner face of the plasma membrane of *H. halobium* R$_1$mR (Nanba and Mukohata, 1987). The ATPase was about (300-) 320 kDa, which was likely composed of two pairs of 86 kDa and 64 (-61) kDa subunits. The ATPase identified in *H. saccharovorum* has the subunit composition very similar to this, i.e. 300 kDa composed of 83 kDa and 60 kDa (Kristjansson et al., 1986). It should be noted that no ATP hydrolyzing enzyme other than this ATPase was active and optimum in 1.5 M Na$_2$SO$_4$ at pH 5.8. The ATPase required Mn^{2+} rather than Mg^{2+}. The apparent Km value for ATP was 1.4 mM and the Ki for ADP (competitive to ATP) was 80 μM. The ATPase was not inhibited by azide or vanadate. Phosphoenzyme formation has not been detected (Post and Mukohata, unpublished results).

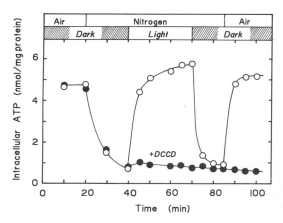

Fig. 1. Changes in the Intracellular ATP Level of Halobacterial Cells. R$_1$mR (bR$^-$, hR$^+$) cells were used in the presence of 2 μM triphenyltin chloride (TPT) which reduced intracellular acidification (Mukohata and Kaji, 1981b). R$_1$ (bR$^+$, hR$^+$) cells showed the similar profile in the absence of TPT. (●); DCCD (100 μM) was added 30 min before illumination. It took about 1 hr for DCCD to inhibit ATP synthesis completely.

Under the conditions optimum and specific to the isolated ATPase, the envelope vesicles showed negligible activity of ATP hydrolysis. However, after an adequate amount of $C_{12}E_9$ (about 0.025%/mg protein) was added, the vesicle suspension became to show the ATPase activity with the pH dependence profile identical to that of the isolated ATPase (Mukohata and Yoshida, 1987b).

The Halobacterial ATP Synthase

Attempts to identify the ATP synthase in halobacteria were succeeded only by correlating the activities of ATP synthesis and hydrolysis after chemical modification of the vesicles (Mukohata and Yoshida, 1987b).

The vesicles were first stuffed with various concentrations of NBD-Cl or NEM. NBD-Cl was used with or without ADP which, when added together with NBD-Cl, partly prevented the inactivation of ATP synthesis in the vesicles (Mukohata and Yoshida, 1987a) and also that of the isolated ATPase (Nanba and Mukohata, 1987). The modifier-stuffed vesicles were incubated for 1 hr in the dark then washed. The vesicles were examined for the activities of ATP synthase (after stuffing the substrates) and ATPase (after solubilized with $C_{12}E_9$). Fig. 3 shows the NBD-Cl-induced inactivation patterns of these two opposing activities. In the absence of protective ADP, more inactivation occurred at higher concentrations of NBD-Cl. In the presence of 3 mM ADP, stimulation, instead of inactivation, was observed at lower NBD-Cl

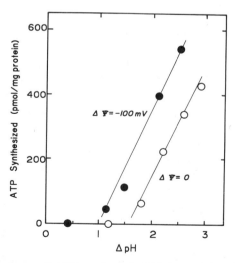

Fig. 2. The Dependence of ATP synthesis in Halobacterial Vesicles on the pH Difference Given by the Base–Acid Transition. The intravesicular pH was fixed at 6.8 and the external pH was changed to give different size of ΔpH in the dark ($\Delta\psi = 0$) or in the light ($\Delta\psi = -100$ mV).

concentrations. The activity profiles of both ATP synthase and ATPase were almost identical even in the partial stimulation in the presence of ADP. The NEM-induced inactivation profiles of synthesis and hydrolysis of ATP were also almost identical, including the partial stimulation at lower NEM concentrations (Mukohata and Yoshida, 1987b).

The correlation factors between the activities of synthesis and hydrolysis of ATP in Fig. 3 were 0.986 and 0.984 for the NBD-Cl(-ADP) and the NBD-Cl(+ADP) modified vesicles, respectively. The factor for the NEM modified vesicles was 0.985. These values strongly imply that these two kinds of activities of opposing reactions are ascribed to one enzyme, the ATP synthase. The 320 kDa ATPase is most likely to be the catalytic part of the synthase. In other words, the halobacterial ATP synthase includes at least two pairs of 86 kDa and 64 kDa subunits.

CONCLUDING REMARKS

The ATP synthesizing machinery in the extremely halophilic archaebacteria is now shown to be different from the ordinary ATP synthase, i.e. F_oF_1-ATPase/synthase (Mukohata and Yoshida, 1987b). It should be noted here that the halobacterial ATP synthase is the first exception from the common concept of F_oF_1-ATPase/synthase; F_oF_1-ATPase/synthase is ubiquitously distributed as the sole machinery of the pmf-coupled ATP synthesis in all respiring organisms on our biosphere.

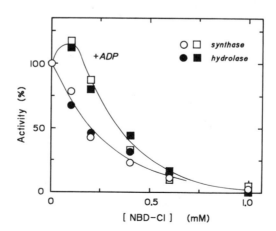

Fig. 3. The Inactivation Profiles of the ATP Synthase and the ATPase in the Vesicles Modified with NBD-Cl in the Presence or Absence of ADP. The vesicles were modified with NBD-Cl in the presence (3 mM) or absence of ADP, and then assayed for ATP synthesis by the base-acid transition or for ATPase after the vesicle components were solubilized by $C_{12}E_9$.

The 320 kDa ATPase component seems to be classified in the family of archaebacterial ATPases, because of similarities in subunit composition, acidic and narrow pH dependence, insensitivity to azide and vanadate and stimulation by sulfite (Mukohata and Yoshida, 1987b).

Furthermore, our preliminary results on the immunological cross-reactions indicated that halobacterial ATPase (the 320 kDa component) related closer to the H^+-ATPase of red beet (*Beta vulgaris*) tonoplasts (Bennett et al., 1984) as well as the ATPase from *Sulfolobus acidocardarius* (a gift of Dr. Masasuke Yoshida, Tokyo Institute of Technology) than F_1-ATPase of spinach chloroplasts and Ca^{2+}-ATPase of sarcoplasmic reticulum (a gift of Dr. Taibo Yamamoto, Osaka University). The enzyme which still synthesizes ATP in halobacteria (and possibly in all archaebacteria) and the H^+-ATPase which now hydrolyzes ATP to pump protons in storage/secretory/regulatory organelles would have the common ancestor (details will be further discussed elsewhere).

REFERENCES

Bennett. A.B., O'Neill, S.D. and Spanswick, R.M. (1984) Plant Physiol. 74, 538-544.

Kamo, N., Muratsugu, M., Hongoh, R. and Kobatake, Y. (1979) J. Membr. Biol. 49, 105-121.

Kristjansson, H., Sadler, M.H. and Hochstein, L.I. (1986) FEMS Microbiol. Rev., 39, 151-157.

Lindley, E.V. and MacDonald, R.E. (1979) Biochem. Biophys. Res. Commun. 88, 491-499.

Matsuno-Yagi, A. and Mukohata, Y. (1977) Biochem. Biophys. Res. Commun. 78, 237-243.

Mukohata, Y., Isoyama, M. and Fuke, A. (1986) J. Biochem. 99, 1-8.

Mukohata, Y., Isoyama, M., Fuke, A., Sugiyama, Y., Ihara, K. and Yoshida, M. (1987) in Perspectives of Biological Energy Transduction (Mukohata. Y. et al., eds.) Academic Press, Tokyo, in press.

Mukohata, Y. and Kaji, Y. (1981a) Arch. Biochem. Biophys. 206, 72-76.

Mukohata, Y. and Kaji. Y. (1981b) Arch. Biochem. Biophys. 208, 615-617.

Mukohata, Y., Matsuno-Yagi, A. and Kaji, Y. (1980) in Saline Environment (Morishita, H. and Masui. M., eds.), pp. 31-37. Business Center for Academic Societies Japan, Tokyo.

Mukohata, Y. and Yoshida, M. (1987a) J. Biochem. 101, 311-318.

Mukohata, Y. and Yoshida, M. (1987b) J. Biochem. 102, in press.

Nanba, T. and Mukohata, Y. (1987) J. Biochem.. 102, in press.

Schobert, B. and Lanyi, J.K. (1982) J. Biol. Chem. 257, 10306-10313.

Taussky, H.H. and Shorr, E. (1953) J. Biol. Chem. 202, 675-685.

THE F$_O$ SECTOR AND PROTON MOTIVE COUPLING

STRUCTURE AND FUNCTION OF MITOCHONDRIAL COUPLING FACTOR B (F_B)

D.R. Sanadi, L. Kantham and R. Raychowdhury

Department of Cell Physiology, Boston Biomedical Research Institute
20 Staniford Street, Boston, MA 02114

F_B is a dithiol protein which stimulates energy-driven reactions in submitochondrial particles depleted of F_B by extraction with ammonia-EDTA (AE). It has been purified extensively and appears as a single band in SDS-PAGE by silver staining (Sanadi, 1982). Its instability, tendency to undergo oligomerization, and low yield have made it a difficult protein to study.

In this communication three aspects of F_B will be discussed: (a) functional role of F_o in H^+-ATPase, (b) evidence for F_B as a subunit of H^+-ATPase and (c) preliminary results on its structure.

Two experimental approaches have been useful in defining the specific role of F_B in the ATP synthase (H^+-ATPase): Isolated F_B activity in stimulating energy-linked reactions is blocked by Cd^{2+}, and the inhibition is not prevented by excess monothiols but is prevented and partially reversed by dithiol compounds. Thus, in the presence of a 5 to 10-fold excess of monothiols over Cd^{2+}, the inhibition becomes relatively specific to dithiol-containing F_B. The Cd^{2+} which is bound tightly to F_B under these conditions can be followed in SDS-PAGE (Sanadi et al., 1985). The second approach involves classical resolution/reconstitution to identify the simplest system in which F_B participates.

INHIBITION OF H^+-ATPase AND F_o ACTIVITY BY Cd^{2+}

Fig. 1 shows that the stimulation of P_i-ATP exchange activity of F_B-depleted AE-particles by added purified F_B is blocked by Cd^{2+} (Fig. 1). The inhibition is relieved in the presence of a 2-fold excess of dithiothreitol but persists in the presence

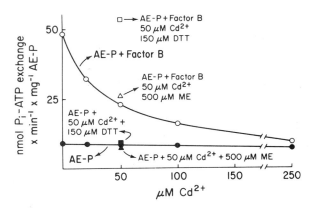

Fig.1. Inhibition of F_B Activity by Cd^{2+} and Its Reversal by Dithiothreitol. Aliquots of 0.5 mg AE particles and/or F_B were incubated with Cd^{2+} in 50 μl for 5 min on ice before assay. Where shown dithiothreitol or 2-mercapto-ethanol was also added to the incubation mixture (Sanadi, 1982).

of even a 10-fold excess of 2-mercaptoethanol. McEnery et al. (1984) have also seen inhibition of rat liver H^+-ATPase activity by Cd^{2+}. Under similar experimental conditions, generation of membrane potential by ATP in the purified, highly active, vesicular H^+-ATPase is also reversibly blocked by Cd^{2+} (Fig. 2) (Pringle and Sanadi, 1984). The membrane potential was monitored by the absorbance change of the voltage sensitive dye, oxonol VI.

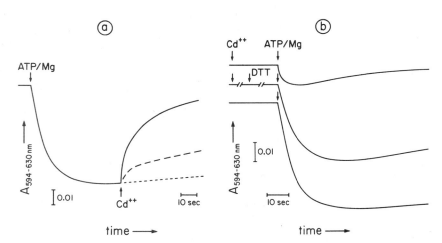

Fig. 2. Inhibition of Oxonol VI Absorbance Changes in Energized H^+-ATPase Vesicles by Cd^{2+}. Vesicles (50 μg F_o-F_1/ml) were energized with ATP/Mg^{2+}. (......) Control with no Cd^{2+}. (---) DTT (500 μM) was added before energization and then Cd^{2+} was added (Pringle and Sanadi, 1984).

152

The specificity of the Cd^{2+} effect was examined with the use of $^{109}Cd^{2+}$, which was added to F_B and to H^+-ATPase in the presence of excess 2-mercaptoethanol, and SDS-PAGE was carried out. The Cd^{2+} labeled protein from the H^+-ATPase comigrates with Cd^{2+} labeled F_B (Fig. 3) (Kantham et al., 1984; Sanadi et al., 1985). The activity peaks coincide almost exactly showing that Cd^{2+} binds to F_B in H^+-ATPase and to nothing else.

Since F_1-ATPase activity is not inhibited by Cd^{2+}, F_o proteoliposomes were examined next. Passive H^+ conductance driven by a K^+ diffusion potential in F_o proteoliposomes is inhibited by Cd^{2+} (Fig. 4) (Sanadi et al., 1984) and, as expected, by oligomycin or dicyclohexylcarbodiimide. DTT affords protection against inhibition by Cd^{2+} but 2-mercaptoethanol does not. As in the case of H^+-ATPase, the protein labeled by Cd^{2+} is recovered as a sharp peak with the M_r of F_B (Fig. 5).

RESOLUTION OF F_B FROM F_o AND ITS RECONSTITUTION

Submitochondrial particles can be depleted of F_B by repeated extraction with ammonia-EDTA (Huang et al., 1987). After three extractions, the particle activity in ATP-driven NAD^+ reduction is stimulated 8-fold by addition of F_B. H^+-ATPase made from these partially F_B-depleted AE-particles contain a significant amount of membrane bound F_B which is further reduced during the preparation of F_o by treatment with NaBr. The presence of traces of F_B in the F_o is shown by the curvature

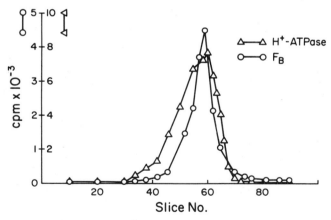

Fig. 3. Comigration of $^{109}Cd^{2+}$-labeled F_B with $^{109}Cd^{2+}$-labeled Protein of H^+-ATPase. The SDS-PAGE was carried out with 10% polyacrylamide gels in 50 mM phosphate (pH 7.0) as running buffer. The gel was sliced and the slices counted. The labeling of the protein was carried out in the presence of a 10-fold excess of 2-mercaptoethanol over Cd^{2+} in order to make the labeling specific to juxtaposed dithiol groups (Sanadi et al., 1985).

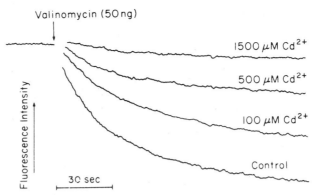

Fig. 4. Inhibition of Passive H$^+$ Conduction of F$_o$-proteoliposomes by Cd^{2+}. Fo was incorporated into liposomes in the presence of 0.2 M KCl and external K$^+$ removed by sedimentation of the vesicles. H$^+$ conduction was monitored by the quenching 9-aminoacridine fluorescence (Sanadi et al., 1985).

of the Ouchterlony precipitin lines (Fig. 7) (Joshi et al. 1979). This extensively depleted F$_o$ shows low P$_i$-ATP exchange activity on incorporation of F$_1$ and the activity is stimulated 4- to 5- fold on addition of purified F$_B$ (Table I). If the F$_o$ is

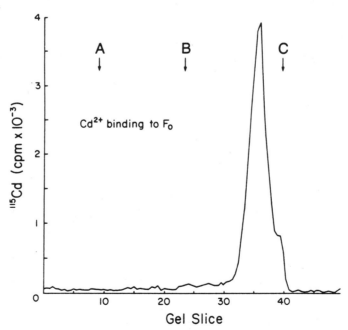

Fig. 5. ^{109}Cd^{2+} Binding to a Specific Protein F$_o$-proteoliposomes. F$_o$ was incubated with 40 μM ^{115}Cd and 200 μM 2-mercaptoethanol and then SDS-PAGE carried out as in Fig. 3. (A) bovine serum albumin, (B) carbonic anhydrase, (C) cytochrome c. The radiolabeled protein had the same M$_r$ as F$_B$ (Sanadi et al., 1984).

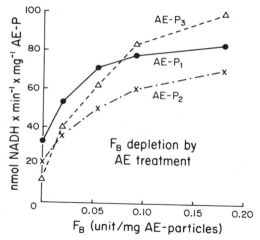

Fig. 6. Progressive Depletion of F_B Reversibly from Submitochondrial Particles by Washing with Ammonia-EDTA. The SMP were washed and reactivation of P_i-ATP exchange activity determined as described in Joshi et al. (1979).

treated with N-ethylmadeimide, the residual activity is completely lost, and F_B becomes an absolute requirement for P_i-ATP exchange activity. However, F_B is not needed for the oligomycin-sensitive ATPase activity; F_1 alone fully restores it, showing that F_B is not needed for binding of F_1 to F_o, but is necessary to convert an OSATPase to an energy conserving system.

Since F_B-depleted F_o failed to show H^+ conductance even after reconstitution with F_B, the involvement of F_B in the reaction was studied in the more intact

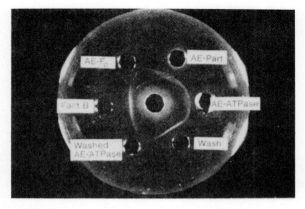

Fig. 7. Detection of F_B in Partially F_B-Depleted Preparations by Ouchterlony Immunodiffusion (Joshi et al., 1979).

Table I. Effects of F_B and Oligomycin on P_i-ATP Exchange and OSATPase Activity of F_o-F_1 Reconstituted from F_1 and F_B-Depleted F_o.

	P_i-ATP Exchange (nmol x min^{-1} x mg^{-1})		OSATPase (mol x min^{-1} x mg^{-1})	
	$-F_B$	$+F_B$	$-$oligo	$+$oligo
AE-F_o	0	0	0	0
AE-F_o + F_1	32	148	0.60	0.13
AE-F_o (NEM) + F_1	0	74	0.61	0.16

Experimental details are described in Joshi et al. (1979).

particulate system. F_1-depleted AE-particles (AEBrP) were incorporated into K^+-loaded liposomes by the freeze-thaw sonication procedure, together with and without added F_B. AEBrP activity in passive H^+ conduction was significant and it was stimulated about 100% in the presence of F_B (Fig. 8) (Huang et al., 1987). This stimulation is far less than that expected from the 8-fold stimulation of reversed electron flow activity of the parent AE-particles by F_B. However, H^+ conductance was inhibited significantly by oligomycin (data not shown) showing the specificity of the conductance.

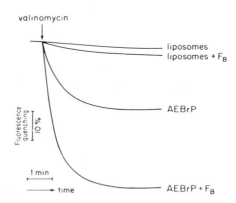

Fig. 8. Stimulation of Passive H^+ Conduction in Proteoliposomes Containing F_1-Depleted AE-Particles (AEBrP) by F_B. The incorporation of the particles into K^+-loaded liposomes was done by the freeze-thaw-sonication procedure and H^+ conduction measured by the quenching of 9-aminoacridine fluorescence. See Huang et al. (1987) for details. F_B incorporation was simultaneous with the preparation of proteoliposomes.

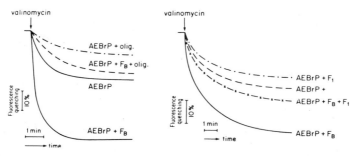

Fig. 9.   Absolute Requirement for F_B in H^+ Conduction in 4-Vinylpyridine Treated AEBrP (AEVpBrP).   AEVpBrP was incorporated with or without added F_B into proteoliposomes and H^+ conduction measured as in Fig. 8.   See Huang et al. (1987) for details.

In order to reduce the H^+ conductance of AEBrP, the AE-particles were first reacted with the -SH reagent, 4-vinylpyridine and then with NaBr to remove F_1. The resulting particle, AEVpBrP, showed absolute dependence on added F_B for H^+ conductivity (Fig. 9) (Huang et al., 1987). These experiments complete the evidence to establish unequivocally that F_B is a functional component of the H^+ channel. If it is inhibited or depleted, the F_0 proton channel is no longer active.

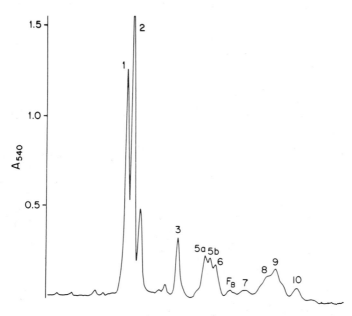

Fig. 10.   SDS-gel Scan of Highly Active H^+-ATPase.   Electrophoresis was according to Weber and Osborn (1969). F_B peak was identified by comigration with highly purified F_B.

Table II. Purification of H^+-ATPase and Enrichment of F_1 and F_B Content.

	Exchange (nmol x min^{-1} x mg^{-1})	F_1 (nmol x mg^{-1})	F_B (nmol x mg^{-1})
ETP	150	0.24	0.27
Lysolecithin extract	400	-	-
H^+-ATPase	1200	2.0	2.1
Enrichment	8.0	8.3	7.8

See Sanadi et al. (1985).

IS F_B A SUBUNIT OF F_o-F_1?

In 1982 we reported the isolation of a highly active H^+-ATPase from bovine heart SMP by extraction with lysolecithin (Hughes et al., 1982). Its SDS gel profile shows 12 or 13 bands (Fig. 10). The peak following that of α, β of F_1 is not seen in some preparations that are still active and hence is considered a contaminant. The F_B peak has been identified by comigration with authentic F_B. The activity of this preparation is over 1500 nmols P_i exchanged x min^{-1} x mg^{-1} protein. The hog heart H^+-ATPase made by the same procedure had even higher activity (Penin et al., 1982). During the preparation, the activity as well as F_1 and F_B content of our preparation all increase about 8-fold (Table II). The exchange activity is not increased by added F_1 or F_B showing the presence of a full complement of these coupling factors. Other preparations reported in the literature, made with the use of ionic detergents and salt precipitation, have only about 10% of the activity observed with the lysolecithin preparation, one example being Tzagoloff's OSATPase which has an activity of about 50 nmoles x min^{-1} x mg^{-1} which increases to 200 after incorporation into liposomes (Table III). This activity is more than doubled on addition of F_B, showing significant loss of F_B during preparation.

In western blots of membrane bound F_B using a monoclonal antibody, the F_B band can be seen but is faint compared to purified F_B, presumably since the epitope is not fully exposed under these conditions (Fig. 11).

The F_B content of ETPH and the H^+-ATPase have been determined by three methods: labeled NEM binding, ELISA and $^{109}Cd^{2+}$ binding. The H^+-ATPase has close to 2.0 nmoles F_B/mg protein which is stoichiometric with the F_1 in the preparation (Table IV) (Sanadi et al., 1985).

Table III. Effect of F_B on P_i-ATP Exchange Activity of ATP Synthase Preparation.

	P$_i$-ATP Exchange (nmol x min^{-1} x mg^{-1})		Reference
	$-F_B$	$+F_B$	
OSATPase[1]	200	450	Joshi et al. (1979)
H$^+$-ATPase[2]	550	570	Sanadi et al. (1979)

[1]OSATPase was prepared as described by Tzagoloff et al. (1968). Its specific activity in exchange was 50 and increased to 200 on incorporation into asolectin liposomes.
[2]Lysolecithin extract after first differential sedimentation. The assays were as in Joshi et al. (1979).

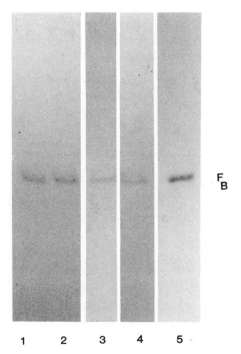

Fig. 11. Reactivity of F_B from Different Membrane-Bound Preparations to Monoclonal Antibody. 1. Mitochondria. 2. ETPH. 3. AE-particles. 4. H$^+$-ATPase. 5. F$_B$. Samples 1 to 4 were 80 μg protein and F_B was 0.013 unit (Joshi et al., 1985).

```
      1                      5                        10
    Phe - Trp- Gly - Trp -Leu- Asn- Ala - Val - Phe - Asn
     F     W     G     W     L     N     A     V     F     N

     11                    15                          20
    Lys - Val - Asp- His - Asp- Arg - Ile - Arg- Asp- Val
     K     V     D     H     D     R     I     R     D     V

     21                    25                          30
    Gly - Pro - Asp- Arg - Ala - Ala - Ser- Glu - Trp - Leu
     G     P     D     R     A     A     S     E     W     L

     31                    35                          40
    Leu - Arg- Gly - Gly - Ala - Met - Val - Arg - Tyr - His
     L     R     G     G     A     M     V     R     Y     H

     41                    45                          50
    Gly - Gln - Gln - Arg - Trp - Gln - Lys - Asp - Tyr - Asn
     G     Q    [Q]    R     W    [Q]    K     D     Y     N

     51                    55
    His - Leu - Pro - Thr  Gly
     H     L     P     T     G
```

Fig. 12. NH$_2$ - Terminal Amino Acid Sequences of F$_B$.

These and other published results establish F$_B$ as a subunit of F$_o$-F$_1$, in particular of the F$_o$ segment.

PRIMARY STRUCTURE OF F$_B$

In order to understand the molecular mechanism of F$_B$ action, the next step is to determine its structure. The NH$_2$-terminal sequence of F$_B$ has been studied by Edman degradation in a gas phase sequencer. The 55 amino acid sequence shown in Fig. 12 is strongly hydrophilic, containing 17 charged residues. The hydropathy profile (Fig. 13) shows two major hydrophilic regions with two short mildly hydrophobic stretches. In a search carried out with the BIONET data base, no homology with the E. coli unc operon, OSCP or F$_6$ was detected. Further sequencing of peptides derived from F$_B$ is underway, as well as determination of primary structure by molecular cloning of F$_B$.

REFERENCES

Huang, Y., Kantham, L. and Sanadi, D.R. (1987) J. Biol. Chem. 262, 3007-3010.

Hughes, J.B., Joshi, S., Torok, K. and Sanadi, D.R. (1982) J. Biomemb. Bioenerg. 14, 287-295.

Joshi, S., Hughes, J.B., Shaik, F. and Sanadi, D.R. (1979) J. Biol. Chem. 254, 10145-10152.

Joshi, S., Kantham, L., Kaplay. S.S. and Sanadi, D.R. (1985) FEBS Letters 179, 143-147.

Kantham, B.C.L., Hughes, J.B., Pringle, M.J. and Sanadi, D.R. (1984) J. Biol. Chem. 259, 10627-10626.

McEnery, M.W., Buhle, E.L. Jr., Aeli, U. and Pedersen, P.L. (1984) J. Biol. Chem. 259, 4642-4651.

Penin, F., Godinot, C., Conte, J. and Gautheron, D.C. (1982) Biochim. Biophys. Acta 679, 198-209.

Pringle, M.J. and Sanadi, D.R. (1984) Membr. Biochem. 5, 225-241.

Sanadi, D.R., Joshi, S. and Shaikh, F.M. (1977) in The Molecular Biology of

Membranes (S. Fleisher, ed.), pp. 263-272. Plenum Press, New York.

Sanadi, D.R. (1982) Biochim. Biophys. Acta 683, 39-56.

Sanadi, D.R., Pringle, M., Kantham, L., Hughes, J.B. and Srivastava, A. (1984) Proc. Natl. Acad. Sci., USA 81, 1371-1374.

Sanadi, D.R., Joshi, S., Huang, Y., Kantham, L., Pringle, M. and Hughes, J.B. (1985) in Achievements and Perspectives of Mitochondrial Research, Vol. 1, Bioenergetics (E. Quagliariello et al. eds.), pp. 269-277. Elsevier Science Publishers.

Tzagoloff, A., Byington, K.H. and MacLennon, D.H. (1968) J. Biol. Chem. 243, 2405

SUBUNIT ARRANGEMENT IN BOVINE MITOCHONDRIAL H⁺-ATPASE

Saroj Joshi and Robert Burrows

Boston Biomedical Research Institute, Department of Cell Physiology
20 Staniford Street, Boston, Mass. 02114

The ATP-synthase or H^+-ATPase of mitochondria is a multisubunit enzyme that couples the energy of a transmembrane proton gradient to the reversible synthesis of ATP (Pedersen, 1983; Hatefi, 1985). The enzyme is comprised of two sectors: the extrinsic sector F_1 carries the catalytic centers and the membrane-intercalated sector F_0 constitutes the proton conduction pathway through the membrane. These two sectors are believed to be connected by another morphologically distinct entity known as the stalk (Soper et al., 1979). The F_1-ATPase from all known sources contains five nonidentical subunits termed α, β, γ, δ and ϵ in a stoichiometry of 3:3:1:1:1 respectively (Wakabayashi et al., 1977). The composition of F_0 is species-dependent. The *E. coli* F_0 (the simplest so far defined) contains three subunits designated *a*, *b* and *c* (Senior and Wise, 1983). Mitohondrial F_0 is considerably more complex and contains at least thirteen subunits. Two of these, namely oligomycin-sensitivity-conferring protein (OSCP) and coupling factor 6 (F_6), are considered to constitute the stalk segment that separates F_1 from the membrane segment (Hatefi, 1985). The organization of the stalk subunits, and how they interact with other F_0 subunits or subunits of F_1 is not clear. However, to understand the functional roles of the polypeptides which comprise a multisubunit complex, a knowledge of their subunit arrangement is essential. A plausible arrangement of subunits in F_1 has been deduced from cross-linking experiments and from X-Ray diffraction studies, but little is known about the organization of subunits belonging to the F_0 region.

In the present study, the structural associations between polypeptides of F_0 were investigated using cleavable cross-linking reagents varying from zero to 11 Å in length. The cross-linked products were analyzed by one and two-dimensional gels followed by western blotting against subunit-specific antisera. The accessibility of subunits of F_0 to the external environment was investigated by following the relative

Molecular Structure, Function, and Assembly of the ATP Synthases
Edited by Sangkot Marzuki
Plenum Press, New York

degradation of these subunits as a result of treating F_O with the membrane-impermeant reagent trypsin. It was assumed that subunits belonging to the stalk should be more accessible than those buried in the lipid bilayer. A consideration of the data obtained from such studies and from those in our recent publication on the topology of OSCP and F_6 (Joshi et al., 1986) has led us to revise a previously proposed model (Hamamoto and Kagawa, 1985) for the arrangement of subunits in the mitochondrial H^+-ATPase.

RESULTS AND DISCUSSION

The H^+-ATPase used in the present investigation was isolated by a modification (Joshi et al., 1986) of the procedure of Hughes et al. (1982). The preparation had a high (1400 U per mg) ATP-P_i exchange and relatively low (5-7 U per mg) ATPase activity that was only marginally stimulated by added lipid or coupling factors. The latter feature was taken to indicate that our isolation procedure has allowed retention of the stoichiometry and subunit organization observed in undissociated mitochondria. This was regarded as an important criterion for selecting this preparation for structural studies. The ATPase complex showed 13-15 major bands in highly-resolving dodecyl sulfate gel electrophoresis (SDS-PAGE) (Fig. 1). Five of these pertained to the known subunits of F_1. Two polypeptides identified as subunit 6 (ATPase 6) and subunit 9

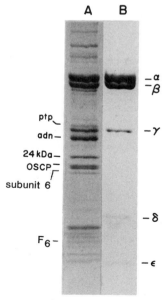

Fig. 1. SDS-PAGE Profile of H^+-ATPase and F_1-ATPase on 12-20% Linear Polyacrylamide Gels.

(DCCD-binding protein), are homologous to *E. coli* subunits *a* and *c* respectively (Senior and Wise, 1983). The identity of subunit 6 was established by western blotting H^+-ATPase against antiserum to a synthetic peptide modeled on the C-terminal sequence of subunit 6. Four further bands, representing oligomycin-sensitivity-conferring protein (OSCP; $M_r = 20,967$), coupling factor 6 (F_6; $M_r = 9,006$), coupling factor B (F_B; $M_r = 15,000$) and subunit 8 (A6L gene product; $M_r = 7,965$), are unique to mitochondrial H^+-ATPase. In addition to these were present two prominent polypeptides of apparent molecular mass of 24 and 20 kDa. Besides these thirteen proteins that were always present in our H^+-ATPase preparation, two polypeptides with apparent molecular masses of 32 and 29 kDa (adenine nucleotide translocase) were occasionally observed.

OSCP and F_6 are required for oligomycin-sensitive ATPase and are believed to constitute the stalk portion between the F_1 and F_o moieties (Hatefi, 1985). Factor B is required for H^+-translocation (Huang et al., 1987). The exact function of the A6L gene product is not known although it is similar in some respects to the AAP1 protein of yeast mitochondrial H^+-ATPase which is required for correct assembly of the complex (Velours et al., 1984). As for the 20 and 24 kDa proteins, polypeptides of this size have been consistently seen in most of the H^+-ATPase preparations derived from bovine heart (Serrano et al., 1976; Galante et al., 1979), rat liver (McEnery et al., 1984), and yeast (Hadikusumo et al., 1984) mitochondria. It is not yet clear whether these two proteins are bona-fide subunits of H^+-ATPase although the 24 kDa protein is reported to be homologous to subunit *b* of bacterial F_o (Fearnley and Walker, 1986; Walker et al., 1987).

The subunit organization in the mitochondrial F_1-F_o was investigated by two different approaches. Firstly, the accessibility of membrane-bound subunits was examined by subjecting preparations of the H^+-ATPase to the impermeant protease, trypsin. Results presented in Table I show that the ATP-P_i exchange activity of native H^+-ATPase (column 2) was rather resistant to trypsin while that of the ATPase reconstituted with trypsin treated-F_o and untreated F_1 (column 3) was more readily inactivated. As seen in Fig. 2, western blot analysis of trypsin-treated F_o showed that degradation of F_o subunits was both time-dependent and concentration-dependent (Joshi et al., 1986). Degradation of OSCP was found to be the most rapid and seemed to parallel the loss of exchange activity (Joshi et al., 1986). During this time course, there was no loss in the staining intensity of F_6, subunit 6, or the 24 kDa protein. The overall order of degradation of F_o subunits was OSCP > 24 kDa protein $\geq F_6$ > subunit 6 (Fig. 2). The structure of undissociated H^+-ATPase was found to be relatively resistant to the action of trypsin. These results suggest to us that in the intact H^+-ATPase, subunits of F_o are essentially shielded by the F_1 subunits. Similarly, the slower

Table I. Effect of Trypsin on the ATP-Pi Exchange Activity

Trypsin/protein (w/w)	Percentage ATP-P_i exchange	
	F_1-F_o	$F_1F_o^*$
	100	100
1:1670	92	97
1:500	96	92
1:167	91	57
1:50	78	20

Aliquots of H^+-ATPase (F_1-F_o) or F_o were diluted to 2.0 mg/ml and incubated with the indicated levels of TPCK-trypsin for 2 minutes at 30°C. The treated F_o samples (indicated by an asterisk) were reconstituted with untreated F_1, and the reconstituted enzyme and the control F_1-F_o samples were assayed for ATP-P_i exchange activity.

degradation of membrane-bound F_6 relative to OSCP might be related to its protection by OSCP or by other subunits belonging to F_o. Although the primary structure of the 24 kDa protein is not known, its degradation by trypsin indicates that the protein contains hydrophilic segments. On the other hand, the limited degradation of subunit 6 by trypsin is consistent both with its low lysine and arginine content as well as its lipophilic nature (Fearnley and Walker, 1986). In summary then, the inaccessibility of membrane-bound proteins to trypsin in intact F_1-F_o suggests that subunits of F_1 are shielding F_o subunits (Joshi et al., 1986) and that OSCP, F_6 and the 24 kDa protein are all located at the interface between the F_1 and F_o moieties.

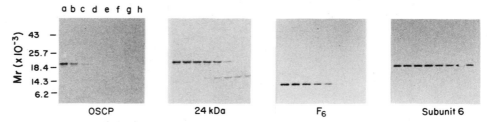

Fig. 2. Western Blot Pattern of Trypsin-Treated F_o. Aliquots of F_o were treated with trypsin at 30°C for (a) 0, (b) 2, (c) 5, (d) 10, (e) 15, (f) 30, (g) 60, and (h) 120 min. Following SDS-PAGE and electrotransfer, the replicas were allowed to react with the indicated antisera.

Secondly, subunit associations in F_o were investigated by using cleavable bifunctional cross-linking reagents such as copper-o-phenanthroline (Cu-o-P; zero length), dithiobissuccinimidyl propionate (DSP; 11 Å), or disuccinimidyl tartrate (DST; 6 Å). Cross-linked samples were analyzed by highly resolving one- and two-dimensional gels, followed by staining with silver or western blotting against a number of subunit-specific antisera.

Treatment of the H^+-ATPase with Cu-o-P resulted in a concentration-dependent formation of a 45 kDa product as revealed by SDS-PAGE (Torok and Joshi, 1985). Following cleavage, the cross-linked product gave rise to a 21 and a 24 kDa component (Fig. 3Ac). The 21 kDa component was subsequently identified as OSCP by western blotting (Fig. 3A a,c; 3B a-c). These observations indicate that the 24 kDa protein and OSCP are intimately associated in our H^+-ATPase preparation.

Treatment of the enzyme complex with DSP and DST resulted in the formation of several cross-links (Table II; Fig. 4-6). Among the cross-links involving F_o subunits were products with apparent molecular masses of 15-47 kDa. OSCP was found to interact with the 24 kDa protein to form a 45 kda product (Fig. 4 G-L; Fig. 6 spots 5 a,b). This is consistent with our previous Cu-o-P data. In addition to OSCP, the 24 kDa protein also formed links with the 20 kDa protein (Fig. 6 spots 4 a,b) and F_6 (Fig. 6 spots 6 a,b) to give rise to products with apparent molecular masses of 47 (Fig. 4 J-L) and 33 kDa (Fig. 4 A-C; J-L) respectively. The F_6 protein was found to participate in the formation of two cross-linked products (33 and 30 kDa) (Fig. 4 A-C) that migrated either with (30 kDa) or slightly behind (33 kDa) the γ subunit of F_1 in SDS-PAGE. Separation of these two products on SDS gels was achieved by allowing proteins smaller than 20 kDa to run off the bottom of the separation gel. Segments of gel containing the two F_6 cross-linked products were excised and re-electrophoresed under reducing conditions. Data presented in Fig. 5 demonstrate that the 33 kDa product was formed by F_6 and the 24 kDa protein (Fig. 5 lane A) and the 30 kDa product F_6 and the 20 kDa protein (Fig. 5 lane B). That the 20 kDa protein

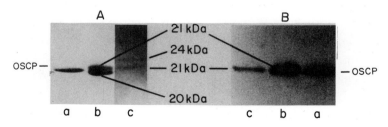

Fig. 3. SDS–PAGE Profile (A) and Immunoblot Pattern (B) of the 45 kDa Product. Antiserum to OSCP was used for immunoblotting.

Table II. Molecular Weights of Crosslinked Species.

	Observed Molecular Mass	Proposed partners
F_o-F_o crosslinks	15 kDa	F_6-8 kDa
	30	F_6-20 kDa
	33	F_6-24 kDa
	45	OSCP-24 kDa
	47	20 kDa-24 kDa
F_1-F_1 crosslinks	41 kDa	$\epsilon\gamma$
	48	$\delta\gamma$
F_1-F_o crosslinks	43 kDa	F_6-γ
	66	F_6-α
	76	OSCP-α/β
	76	24 kDa-β
	80	OSCP-α

The molecular weights of the crosslinked species were estimated from a standard curve constructed using the mobilities of unmodified subunits of H^+-ATPase (Fig. 4 and 6). The respective molecular weights of the β, γ, δ, OSCP, and F_6 were taken to be 51595, 30141, 15065 (Walker et al., 1985); 20967 (Ovchinnikov et al., 1984); and 9006 (Fang et al., 1984).

participating in this cross-link is different from OSCP, was established by the lack of reactivity of this band to OSCP-specific antiserum (Fig. 4 G-I; Fig. 6 spots 7 a,b). In addition to these cross-links, F_6 was found to form another product of 15 kDa (Fig. 4 A-C; Fig. 6 spots 8 a,b). The identity of the other participant was not established although its size is approximately that expected for the A6L gene product (Fearnley and Walker, 1986).

Among the cross-links involving subunits of F_1-ATPase, two products were identified in 2-dimensional diagonal gels. The 48 kDa product was found to arise from association of the γ and δ subunits. The 41 kDa product was comprised of γ and a protein that migrated with a mobility corresponding to the ϵ subunit in H^+-ATPase.

Products involving subunits of F_o and F_1 migrated in SDS-PAGE with apparent molecular masses corresponding to 43-80 kDa. OSCP was found to associate with

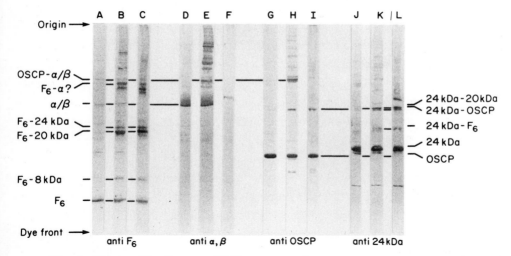

Fig. 4. Western Blot Pattern of DSP-Treated H$^+$-ATPase and F$_o$ after One Dimensional SDS-PAGE. Electrotransferred replicas were allowed to react with the indicated antisera. A,D,G,J - H$^+$-ATPase; B,E,H,K - DSP-treated H$^+$-ATPase; C,F,I,L - DSP-treated F$_o$.

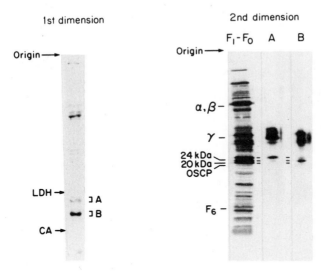

Fig. 5. Identification of Cross-Linked Partners of F$_6$. Aliquots of DSP-treated H$^+$-ATPase were electrophoresed on 15% polyacrylamide gels with prestained lactate dehydrogenase (LDH) and carbonic anhydrase (CA) as markers. Electrophoresis was continued until the CA marker reached the bottom of the separation gel. Segments A and B, corresponding to the anti-F$_6$ reactive cross-linked products were excised and re-electrophoresed in the presence of reducing reagent, followed by staining with silver.

169

both the α and β subunits of F_1 (Fig. 4 D-H; Fig. 6 spots 1 a,b). These interactions have also been observed using zero-length cross-linkers (Dupuis et al., 1985; Archinard et al., 1986) or another 11 Å long cross-linker, namely dimethyl suberimidate (Archinard et al., 1986). F_6 formed cross-links with the α (Fig. 4 B,E; Fig. 6 spots 2 a,b) and γ (Fig. 6 spots 3 a,b) subunits of F_1 giving rise to a <u>66</u> and a <u>43 kDa</u> product respectively. The 24 kDa protein also interacted with the α and/or β subunit of F_1. These associations between subunits of F_1 with F_6 and the 24 kDa protein have not been previously reported.

Our data using cross-linking reagents demonstrate interactions of OSCP and F_6 (the two so-called stalk proteins) with subunits of F_1 as well as with other membrane-bound subunits of F_0. The same holds true for the 24 kDa protein. This is consistent with our trypsinization data and suggests that all three proteins are present at the interface between the F_1 and F_0 moieties, and together constitute the stalk region.

Fig. 7 is a schematic representation of our present ideas of the subunit structure of mitochondrial H$^+$-ATPase. The subunits are shown as being organized into three

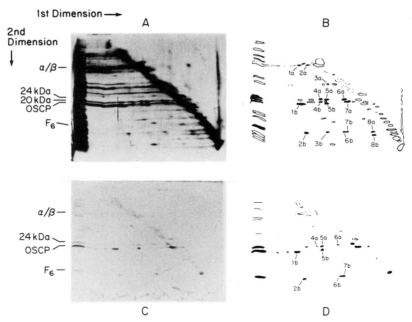

Fig. 6. SDS–PAGE Profile and Western Blot Pattern of DSP-Treated H$^+$-ATPase in Diagonal Gel Electrophoresis. The developed electropherograms were silver stained (A) or electroblotted (C) and sequentially stained with antisera to F_6, OSCP and the 24 kDa protein. B and D are schematics of A and C respectively. The cross-linked proteins migrated below the diagonal. The putative partners of a cross-link migrated along the same vertical and were denoted by same number but different letters.

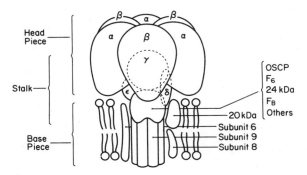

Fig. 7. Subunit Structure of Mitochondrial H^+-ATPase as Modified from Fig. 3 of Hamamoto and Kagawa (1985).

regions designated as the head piece, the stalk, and the base piece. The head piece contains subunits of F_1, the base piece contains subunits of F_0 and the stalk region is composed of subunits of F_1 as well as F_0. The location of γ, δ and ϵ subunits of F_1 in the present diagram is consistent with the observed cross-links of γ with F_6, the δ and possibly ϵ subunit. The α and β subunits are shown as extending down from the head to the base piece in order to account for the observed shielding of the trypsin-sensitive domains of OSCP, F_6 and the 24 kDa protein from the protease. The stalk region is not completely defined but is shown to contain OSCP, F_6 and the 24 kDa protein since all three form cross-links with proteins belonging to both the head piece (F1) and the base piece (F_0). The 20 kDa protein is shown next to the stalk region, which is consistent with its cross-linking to F_6 and the 24 kDa protein. F_B is also placed in the stalk region although firm evidence supporting its localization is not yet available. Subunits 6 and 8 (A6L gene product) are shown as being in the base piece or lipid bilayer which is consistent with their lipophilic nature as well as with hydropathy profiles as predicted from their primary structure. It may be pointed out that this model does not take into account the size, shape, or stoichiometry of some of the subunits. More precise knowledge of the functional organization of the subunits will require additional information on their stoichiometry and their role in energy coupling.

REFERENCES

Archinard, P., Godinot, C., Comte, J. and Gautheron, D.C. (1986) Biochemistry 25, 3397-3404.

Dupuis, A., Lunardi, J., Issartel, J.P. and Vignais, P.V. (1985) Biochemistry 24, 734-739.

Fang, J.-K., Jacobs, J.W., Kanner, B.I., Racker, E. and Bradshaw, R.A. (1984) Proc. Natl. Acad. Sci. USA 81, 6603-6607.

Fearnley, I.M. and Walker, J.E. (1986) EMBO J. 5, 2003-2008.

Galante, Y.M., Wong, S.Y. and Hatefi, Y. (1979) J. Biol. Chem. 259, 12372-12378.

Hadikusumo, R.G., Hertzog, P.J. and Marzuki, S. (1984) Biochim. Biophys. Acta 765, 258-267.

Hamamoto, T. and Kagawa, Y. (1985) In *The enzymes of biological membranes Vol. 4* (Martonosi, A.N., ed.), pp. 149-176, Plenum Press, N.Y.

Hatefi, Y. (1985) Annu. Rev. Biochem. 54, 1015-1069.

Huang, Y., Kantham, L. and Sanadi, D.R. (1987) J. Biol. Chem. 262, 3007-3010.

Hughes, J., Joshi, S., Torok, K. and Sanadi, D.R. (1982) J. Bioenerg. Biomembr. 14, 287-295.

Joshi, S., Pringle, M.J. and Siber, R. (1986) J. Biol. Chem. 261, 10653-10658.

McEnery, M.W., Buhle, E.L., Jr, Aebi, U. and Pedersen, P.L. (1984) J. Biol. Chem. 259, 4642-4651.

Ovchinnikov, Y.A., Modyanov, N.N., Grinkevich, V.A., Aldanova, N.A., Trubetskaya, O.E., Nazimov, I.V., Hundal, T. and Ernster, L. (1984) FEBS Lett. 166, 19-22.

Pedersen, P.L. (1983) Ann. N.Y. Acad. Sci. 402, 1-20.

Senior, A.E. and Wise, J.G. (1983). J. Membr. Biol. 73, 105-124.

Serrano, R., Kanner, B.I. and Racker, E. (1976) J. Biol. Chem. 251, 2453-2461.

Soper, J.W., Decker, G.L. and Pedersen, P.L. (1979) J. Biol. Chem. 254, 11170-11176.

Torok, K. and Joshi, S. (1985) Eur. J. Biochem. 153, 155-159.

Velours, J., Esparza, M., Hoppe, J., Sebald, W. and Guerin, B. (1984) EMBO J. 3, 207-212.

Wakabayashi, T., Kubota, M., Yoshida, M. and Kagawa, Y. (1977) Mol. Biol. (Mosc) 117, 515-519.

Walker, J.E., Fearnley, I.M., Gay, N.J., Gibson, B.W., Northrop, F.D., Powell S.J., Runswick, M.J., Saraste, M. and Tybulewicz, V.L.J. (1985) J. Mol. Biol. 184, 677-701.

Walker, J.E., Cozens, A.L., Dyer, M.R., Fearnley, I.M., Powell, S.J. and Runswick, M.J. (1987) Biochem. Soc. Trans. 15, 104-106.

INTERACTION OF REGULATORY SUBUNITS WITH THE F_1 SECTOR OF

ATP SYNTHASE IN MITOCHONDRIA

Kunio Tagawa*, Tadao Hashimoto and Yukuo Yoshida

Department of Physiological Chemistry, Medical School
Osaka University, Kita-ku, Osaka 530, Japan

ABSTRACT

Mitochondrial ATP synthase has two stabilizing factors, 9K protein and 15K protein. The 9K protein was found to bind directly to F_1-ATPase in the same way as ATPase inhibitor does. When 9K protein and the inhibitor were added simultaneously to F_1-ATPase, they bound to the enzyme competitively, while when added successively with several minutes between their additions only the one added first bound tightly and it was scarcely replaced by the other added later. The F_1-ATPase-9K protein complex retained about 60% of the ATP-hydrolyzing activity of the free enzyme and this activity was not inhibited by addition of the inhibitor. These characteristic binding features of the inhibitor and 9K protein were also observed with membrane-bound F_1F_0-ATPase unless the other stabilizing factor, 15K protein, was present. In the presence of 15K protein both the inhibitor and 9K protein could bind to F_1F_0-ATPase simultaneously but only one of the two ligands interacted directly with the F_1 sector; exchange with the other ligand could not be achieved by the enzyme alone. These results indicated the existence of an assembly for regulating the activity of mitochondrial ATP synthase consisting of an inhibitor and stabilizing factors; the inhibitor and stabilizing 9K protein are counterparts in regulation of the enzyme activity, their bindings resulting in inactive and active F_1, respectively.

INTRODUCTION

An intrinsic proteinous inhibitor of F_1F_0-ATPase found in mitochondria plays an important role in regulation of oxidative phosphorylation (Pullman and Monroy, 1963; Schwerzmann and Pedersen, 1986). The inhibitor binds to purified F_1-ATPase and membrane-bound F_1F_0-ATPase in a 1:1 molar ratio, forming a completely inactivated complex (Gomez-Fernandez and Harris, 1978; Hashimoto et al., 1981). The complex gradually dissociates *in vitro* when ATP is removed from the medium. However, in mitochondria, the ATP-hydrolyzing activity of ATP synthase is inhibited by the inhibitor even in the absence of ATP, suggesting that there must be factors that stabilize the inactivated complex between the enzyme and inhibitor. In fact, a fraction obtained from mitochondrial extracts was found to stabilize the inhibited membrane-bound enzyme, but not the purified F_1 complex. Therefore, the factors were first called "inactivated F_1F_0-ATPase complex-stabilizing factor" or simply "stabilizing factors" (Hashimoto et al., 1983), and later were separated into two proteins, 9K and 15K proteins (Hashimoto et al., 1984).

In addition to having regulatory functions, it seems very significant for their physiological role that the two stabilizing factors and the inhibitor protein are present in equimolar ratios to F_1-ATPase in membranes and that all these proteins remain in the membrane fraction after extraction of F_1F_0-ATPase with Triton X-100 (Okada et al., 1986). These observations strongly suggested that the inhibitor and the two stabilizing factors are present as an assembly in the mitochondrial membranes, and that this assembly with F_1F_0-ATPase forms the complete structure of ATP synthase on the inner membrane.

In the present paper we studied the direct interactions of the inhibitor protein and the two stabilizing factor proteins with the F_1 and F_0 sectors of ATP synthase. We found that purified F_1-ATPase bound to either the inhibitor or the 9K protein, but not both simultaneously, whereas membrane-bound F_1F_0-ATPase could bind to both the inhibitor and 9K protein with the aid of the 15K protein. On binding to all three proteins the F_1F_0-ATPase complex had either no ATP-hydrolyzing activity or partial activity, depending on the extents of direct interaction of the F_1 sector with the inhibitor and 9K protein, respectively.

EXPERIMENTAL PROCEDURES

Materials

$[^{14}C]$Phenylisothiocyanate (PITC, 30 mCi/mmol) was obtained from Amersham. TSK-gel G 3000 SW was obtained from Toyosoda Co. Ltd. ATPase inhibitor (Hashimoto et al., 1981), F_1-ATPase, 9K and 15K proteins (Hashimoto et al., 1984), and F_1F_0-ATPase (Tzagoloff and Meagher, 1972) were prepared by the reported methods. Proteins were labeled with $[^{14}C]$-PITC as described by Levy and Dawson (1976).

Assay of Binding of Ligands to Enzymes

ATPase was incubated with ^{14}C-labeled proteinous factors in medium containing 100 mM MOPS, pH 6.5, 3 mM ATP and 3 mM $MgSO_4$ to obtain enzyme-ligand complexes. These complexes were separated by gel permeation chromatography on a TSK-gel G 3000 SW column (0.75 x 14 cm) equilibrated with 100 mM MOPS, pH 6.5, containing 0.2% Triton X-100. The amounts of ligands bound to the enzyme were calculated from their radioactivities recovered in the enzyme fraction. ATPase activity was assayed by the method of Hashimoto et al. (1981).

RESULTS

One of the Stabilizing Factors, 9K Protein, Binds to F_1-ATPase in the Same Way as ATPase Inhibitor Does

Previously we reported that neither of the two stabilizing factors interacted directly with F_1-ATPase (Hashimoto et al., 1984). In this previous work we examined the binding only at neutral pH. We found in this work, however, that at slightly acidic pH, 9K protein also bound to F_1-ATPase in the presence of ATP and Mg^{2+} as the ATPase inhibitor does (Hashimoto et al., 1981). The F_1-ATPase-9K complex formed was fairly stable at pH 6.5 and could be separated from the excess 9K protein added by gel permeation chromatography. We measured the amount of 9K protein bound to F_1 enzyme quantitatively, and observed a 1:1 molar ratio of binding at saturated concentrations of 9K protein (Table I). The complex had very similar properties to those of the F_1-inhibitor complex, except that it retained about 60% of the ATP-hydrolysing activity of the free F_1-ATPase (data not shown, cf. Fig. 1 and Table II).

ATPase Inhibitor and 9K Protein Bind Competitively to F_1-ATPase

Table I also shows that the inhibitor and 9K protein bound competitively to F_1-ATPase when added simultaneously to the enzyme. The amount of labeled inhibitor bound decreased with increase in the amount of unlabeled 9K protein added and the decrease was even greater with the reverse combination, reflecting the slightly higher affinity of the inhibitor protein.

Inhibitor Protein Bound to F_1-ATPase was not Replaced by 9K Protein Added Externally, and Vice Versa

The binding kinetics of the inhibitor to F_1-ATPase is very peculiar (Hashimoto

Table I. Competitive Binding of 9K Protein and ATPase Inhibitor to F_1-ATPase.

Addition	AI/F_1	$9K/F_1$
*AI	0.90	
*AI, 9K(10 μg)	0.75	
*AI, 9K(30 μg)	0.58	
*9K		0.88
*9K, AI(10 μg)		0.30
*9K, AI(30 μg)		0.15

Samples of F_1-ATPase (70 μg) were incubated at 20°C for 10 min with 10 μg of ^{14}C-labeled (*) ATPase inhibitor or 9K protein and the indicated amounts of unlabeled ATPase inhibitor or of 9K protein. The amounts of ligands bound to the enzyme were determined as described in "Experimental Procedures"

et al., 1984) and so was that of 9K protein. The two proteins did not show competitive binding when either one was added 10 min before the other (data not shown). Similar results were obtained on the bindings of the inhibitor and 9K protein to the membrane-bound F_1F_0-ATPase unless the other stabilizing factor, 15K protein, was

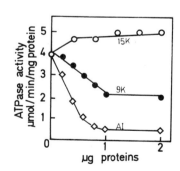

Fig. 1. Effects of 9K Protein, ATPase Inhibitor and 15K Protein on ATPase Activity of Submitochondrial Particles. After the complete activation of bound F_1F_0-ATPase, submitochondrial particles (0.4 mg) were incubated for 10 min in 100 μl of the medium described in the text and their ATPase activities were measured.

Table II. No Exchange of Bound ATPase Inhibitor with External 9K Protein and *Vice Versa*.

Addition		ATPase activity
1st	2nd	(μmol/min/mg protein)
–	–	3.0
AI	–	0.7
9K	–	2.2
AI	9K	0.6
9K	AI	1.2

ATPase inhibitor and 9K protein were added to submitochondrial particles in MOPS buffer with an interval of 10 min between their additions. Ten min after the second addition, the ATPase activity of each sample was measured.

present. Fig. 1 shows the inhibitions of submitochondrial F_1F_0-ATPase by the additions of the inhibitor and the two stabilizing factors. The enzyme was not inhibited by the 15K protein but was partially inhibited by the 9K protein and completely inhibited by the inhibitor, like purified F_1-ATPase. The extent of inhibition by the inhibitor protein was scarcely changed by addition of excess 9K protein 10 min later, and *vice versa* (Table II). These results indicated that the ligand that bound first to the F_1-ATPase could not be exchanged for the other ligand added later.

In the Presence of 15K Protein, 1 mol Each of the Inhibitor and 9K Protein Bound to F_1F_0-ATPase

The 15K protein did not affect the ATP-hydrolyzing activity of F_1F_0-ATPase, but it was found to interact directly with F_1F_0-ATPase, altering the bindings of this enzyme to the inhibitor and 9K protein. As shown in Table III, 15K protein also bound to the membrane-bound enzyme with the same stoichiometry as the inhibitor and 9K protein, although its binding did not require ATP or Mg^{++}. In the absence of 15K protein, the inhibitor and 9K protein bound competitively to the enzyme, as to F_1-ATPase, and the sum of the amounts of these two proteins bound did not exceed 1 mol per mol of enzyme. However, in the presence of 15K protein, the F_1F_0-ATPase complex with the inhibitor or the 9K protein could also bind the other ligand. Thus, on addition of both ^{14}C-labeled proteins together the radioactivity bound to the enzyme was doubled, indicating binding of 1 mol of each protein. The inhibition of

Table III. Binding of One Mole Each of ATPase Inhibitor and 9K Protein to F_1F_0-ATPase.

| Addition | | Radioactivity | Molar ratio | | |
1st	2nd	(counts/10 min)	AI/F_1	$9K/F_1$	$15K/F_1$
*15K	-	2,480			0.8
*AI	-	3,840	0.9		
*9K	-	3,570		0.9	
15K, 9K	*AI	4,000	1.0		
15K, AI	*9K	4,600		1.2	
15K	*AI, *9K	8,400	1.0	1.0	

Various ligands were added as indicated to 0.027 nmol of F_1F_0-ATPase under the same conditions as for Table I except for the presence of a final concentration of 0.6% Triton X-100. The amounts of ligands bound to the enzyme were described in Experimental Procedures.

ATP-hydrolyzing activity was complete within seconds when these three proteins were added simultaneously, whereas it was about 60% when 9K protein was added before the inhibitor, even when excess inhibitor was added later. These results strongly indicated that only one of the two ligands bound to the membrane-bound enzyme interacted directly with the F_1 sector and that this ligand could not be replaced by the other ligand by the enzyme itself.

DISCUSSION

In the present work we found that one of the stabilizing factors of mitochondrial ATP synthase, 9K protein, is a counterpart of the inhibitor protein in regulating this enzyme. When added simultaneously as free forms, the two proteins competed with each other for interaction with the F_1 sector. However, this competition cannot occur in mitochondria, since the two proteins seem to be bound to the membrane with the aid of 15K protein. probably these three proteins form an assembly for regulation of ATP synthase (Hashimoto et al., 1984), which facilitates the binding of either of the two ligands to the F_1 sector and also prevents release of the unbound ligand from the enzyme. The enzyme may exist as either an active or inactive form in mitochondria under physiological conditions, depending on the direct

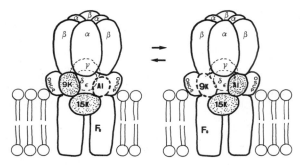

Fig. 2. Model for Alternate Interactions of 9K Protein and ATPase Inhibitor with the F_1 Sector in Mitochondrial ATP Synthase.

interaction of the F_1 sector with 9K protein or the inhibitor, as illustrated in Fig. 2. The enzyme cannot alone reverse the interaction, and so there must be some mechanism for reversible bindings of 9K protein and the inhibitor to the F_1 sector, which probably depends on changes in external conditions, such as changes in electrochemical potential. Actually, dissociation of the membrane-bound F_1-inhibitor complex upon energization of the membrane has been detected by many investigators, as increase of ATP-hydrolyzing activity (Van De Stadt et al., 1973; Tuena De Gomez-Puyou et al., 1980; Schwerzmann and Pedersen, 1981) or release of labeled inhibitor (Klein and Vignais, 1983; Powers et al., 1983). These previous studies have also indicated the existence of several activity states of membrane-bound ATPase, as extensively documented recently (Schwerzmann and Pedersen, 1986). We believe that at least some of these activity states are closely related with the interaction of the inhibitor and 9K protein with the F_1 sector, and can be reasonably interpreted by supposing the existence of a regulatory sector composed of the inhibitor and the two stabilizing factors in mitochondrial ATP synthase.

REFERENCES

Gomez-Fernandez, J.C. and Harris, D.A. (1978) Biochem. J. 176, 967-975.

Hashimoto, T., Negawa, Y. and Tagawa, K. (1981) J. Biochem. 90, 1141-1150.

Hashimoto, T., Yoshida, Y. and Tagawa, K. (1983) J. Biochem. 94, 715-720.

Hashimoto, T., Yoshida, Y. and Tagawa, K. (1984) J. Biochem. 95, 131-136.

Kanazawa, H., Kayano, T., Kiyasu, T. and Futai, M. (1982) Biochem. Biophys. Res. Commun. 105, 1257-1264.

Klein, G., and Vignais, P.J. (1983) J. Bioenerg. Biomemb. 15, 347-362.

Matsubara, H., Hase, T., Hashimoto, T. and Tagawa, K. (1981) J. Biochem. 90, 1159-1165.

Matsubara, H, Inoue, K, Hashimoto, T, Yoshida, Y. and Tagawa, K. (1983) J. Biochem. 94, 315-318.

Okada, Y., Hashimoto, T., Yoshida, Y. and Tagawa, K. (1986) J. Biochem. 99, 251-256.

Powers, J., Crofts, R.L. and Harris, D.A. (1983) Biochim. Biophys. Acta 724, 128-141.

Pullman, M.E. and Monroy, G.C. (1963) J. Biol. Chem. 238, 3762-3769.

Schwerzmann, K. and Pedersen, P.L. (1981) Biochemistry 20, 6305-6311.

Schwerzmann, K. and Pedersen, P.L. (1986) Arch. Biochem. Biophys. 250, 1-18.

Tuena De Gomez-Puyou, M., Nordenbrand, K., Muller, U., Gomez-Puyou, A. and Ernster, L. (1980) Biochim. Biophys. Acta 592, 385-395.

Tzagoloff, A. and Meagher, P. (1972) J. Biol. Chem. 247, 594-603.

Van De Stadt, R.J., De Boer, B.L. and Van Dam, K. (1973) Biochim. Biophys. Acta 292, 338-349.

THE pKa GATE MECHANISM FOR PROTOMOTIVE COUPLING:

EFFECTS OF INTERNAL BUFFERS ON INITIAL RATES OF ATP SYNTHESIS

Evangelos N. Moudrianakis and Robert D. Horner

Department of Biology, Johns Hopkins University
Baltimore, Maryland 21218

INTRODUCTION

The chemiosmotic hypothesis proposed by Mitchell (1966) has maintained that an energy transducing membrane is a semipermeable barrier separating two, osmotically distinct, aqueous spaces, and that the "high energy" intermediate between electron transport and ATP synthesis is the electrochemical gradient of protons consisting of an electrical potential difference ($\Delta\psi$) and a pH difference (ΔpH) between the two spaces. This electrochemical gradient was proposed to be generated by anisotropic electron transport enzymes spanning the otherwise proton-impermeable membrane which transduced oxidation/reduction reactions into transmembrane vectorial proton fluxes. The proton permeability barrier of the biomembrane is interrupted by an anisotropic ATPase which when operating in the forward direction expended the protomotive potential to generate ATP. Not only does this hypothesis predict structures and links them to functions, but it also provides a rational formulation of the thermodynamic properties which govern the rates of the reactions of membrane-bound, energy-transducing enzymatic complexes. This theme of anisotropic, localized, macromolecular "machines" endowed with vectorial functions and linked by a global thermodynamic property, the electrochemical potential, forms the basis of most experiments in bioenergetics today, whether or not the experimental results are consistent with the original formulation of the chemiosmotic hypothesis.

The chemiosmotic hypothesis proposes that the linkage between electron transport and ATP synthesis is established via a transmembrane concentration gradient of protons freely swimming about in a bulk aqueous phase. In opposition, there exist a number of proposals which maintain that individual electron transport complexes directly deliver protons to individual ATP synthetase complexes without those protons first entering the large volume of bulk water (Kell, 1979; Williams, 1978; Theg and Junge, 1983). Since these proposals envision protons traveling along specific structures or pathways, they have been termed "localized proton" models while the protons in bulk solution, according to the chemiosmotic hypothesis, have been termed "delocalized protons". The localized proton models retain the active proton species, not as a hydronium ion in solution, but as a mobile entity being carried along a path within the membrane, within a membrane protein, or on the surface of the membrane with the proviso that the structure keeps the proton from rapid equilibrium with the bulk aqueous phase. This type of model for energy transduction implies that rates of electron transport and ATP synthesis should be directly linked and, moreover, should not be directly dependent on the measurable electrochemical activity gradient between the inner and outer bulk phases.

In between the chemiosmotic model and the localized proton model are hybrid models which incorporate elements of both (Sigalat et al., 1985; Beard and Dilley, 1986; Rottenberg and Steiner-Mordoch, 1986). These models have attempted to account for apparent chemiosmotic behavior under one set of conditions and localized proton behavior under another set of conditions. For instance, the model of Sigalat et al. proposes that in solutions of low osmolarity the inner aqueous space of chloroplasts is large, and consequently chemiosmotic behavior is observed. In solutions of high osmolarity, the model maintains that the inner aqueous space shrinks until the remaining water becomes highly structured, and the system exhibits localized proton behavior.

To distinguish between these models of coupling, a common experimental approach has been to modulate electron transport, or the extent of coupling, or the phosphorylation potential and to measure the resulting changes in both the electrochemical potential and the rates of ATP synthesis. This approach is limited to steady state measurements because most methods which allow estimation of the electrochemical potential are indirect and measure the steady state distribution of mobile ions, which have moved across the membrane in response to a proton gradient or an electrical potential. This steady state approach yields useful information, but it cannot be used to study the initial kinetics of the development of the electrochemical potential, a much more difficult undertaking. The measuring techniques for initial kinetic studies must have response times which are shorter than the processes being measured. For a complete kinetic characterization of the initial development and utilization of the electrochemical potential, one would need to measure the following processes with response times no longer than 2-5 ms: electron transport, proton uptake, membrane potential, internal pH, ATP synthesis, and non-proton ion movements. Such data would provide the best description of the nature of the electrochemical potential and its relation to electron transport and ATP synthesis.

Accordingly, our basic experimental approach has been a kinetic characterization of the initial rates of ATP synthesis in the photophosphorylating membranes of chloroplasts. We influence the electrochemical potential by the introduction of a buffer within the inner aqueous space of the chloroplast in such a way as to delay the development of the transmembrane proton gradient and under conditions where the membrane potential is absent. We have observed that these buffers do not delay the initiation of photosynchronous phosphorylation but do delay the initiation of postillumination phosphorylation. Since neither the chemiosmotic hypothesis nor the various localized proton models can account for both phenomena, these observations have led us to propose that a proton gate exists which partitions proton flow initially between the ATP synthetase complex and the lumen. Thus initial photosynchronous phosphorylation can begin unaffected by a buffer in the lumen, while the development of postillumination or "capacitance" phosphorylation, which depends upon the transmembrane proton gradient, can be delayed by a buffer in the lumen.

METHODS

Preparation of chloroplasts and rapid flow-quench measurements were described by Horner and Moudrianakis (1983, 1986). Subchloroplast particles were prepared according to McCarty (1971) supplementing the isolation medium with 0.1 M N-morpholinoethane sulfonic acid (MES); or 0.1 M N-morpholinopropane sulfonic acid (MOPS); or 0.1 M glycine, 45 mM NaCl. Postillumination phosphorylation (Xe) was assayed according to Jagendorf (1972), and light-induced proton uptake was assayed according to Dilley (1972).

RESULTS

The term "photophosphorylation" refers to ATP synthesis resulting from light-mediated activities of the photosynthetic apparatus, usually in the steady state. Our rapid flow-quench assays allow us to detect a component of photophosphorylation which is not described adequately by the existing nomenclature. This component represents ATP synthesis during the initial period of illumination strictly linked to ongoing electron transport, and is unaffected by agents which affect postillumination phosphorylation. We proposed the term <u>photosynchronous</u> phosphorylation to describe this process (Horner and Moudrianakis, 1986).

The chemiosmotic hypothesis predicts that in the presence of valinomycin and K^+, ATP synthesis should be totally dependent upon the transmembrane proton gradient, and Fig. 1 shows the predicted effects of the two amine buffers imidazole (pKa 6.8) and pyridine (pKa 5.2) on the pH of the inner aqueous space of the chloroplast and on the initiation of ATP synthesis. Taking the minimum transmembrane pH gradient required for ATP synthesis as 2.3 pH units (Jagendorf and Uribe, 1966), the buffer imidazole should delay the drop in the internal pH because its pKa lies between the external pH of 8 and the internal pH required for ATP synthesis, i.e. pH 5.7; however, after a steady state of internal pH 5.3 has been reached (Avron, 1978), imidazole will not greatly extend postillumination ATP synthesis because only 5% of its protons can be used before the threshold (pH 5.7) is reached. Pyridine will not delay the initiation of ATP synthesis as much as imidazole because its maximum buffering range is below the threshold for ATP synthesis. After steady state has been attained it will extend postillumination ATP synthesis much further than imidazole because before the internal pH reaches a value of 5.7 the protonated pyridine will release most of the protons it had captured earlier, and thus further sustain ATP synthesis.

When these predictions were first tested, the rapid flash technique was used to assay initial photophosphorylation, and permeant buffers were reported to have no effect on initial ATP synthesis in one report (Ort et al., 1976) and to delay initial ATP synthesis in two others (Vinkler et al., 1980; Davenport and McCarty, 1980). These apparent differences have been explained now (Horner and Moudrianakis, 1986).

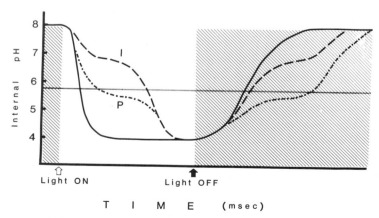

Fig. 1. Predictions of the Chemiosmotic Hypothesis on the Effect of Permeant Buffers on the Internal pH of Chloroplasts in the Presence of Valinomycin and K^+. Imidazole (I); pyridine (P).

Using rapid flow-quench techniques to simulate a rapid flash experiment, we found that there was significant postillumination ATP synthesis after only 100 ms of illumination (Fig. 2A). This implied that any ATP that was <u>considered</u> to have been made during a rapid flash was, in actuality, the combination of ATP made during the light flash *per se* and ATP made in the darkness following the flash. In addition we observed that while the presence of imidazole had small effects on ATP synthesis in the light, it had a large effect on the amount of ATP made after the light (Fig. 2B). It appeared that this permeant buffer affected photosynchronous phosphorylation in a different manner than postillumination (capacitance) phosphorylation.

This property was confirmed when photosynchronous and postillumination phosphorylation were studied in greater detail using the weak amine aniline. The effect of 13 mM aniline on initial photosynchronous phosphorylation at pH 7.2 is shown in Fig. 3. The curves are biphasic with a linear phase beginning at 64 ms and a rapidly increasing phase beginning at 254 ms. It is clear that aniline delays neither phase. The inset shows more clearly that although initial rates with aniline are somewhat lower than the control, the presence of aniline does not delay the formation of the state (electrochemical potential) which drives rapid photophosphorylation. The continuously increasing rates show that after initiation, ATP synthesis is driven by a steadily increasing chemical potential.

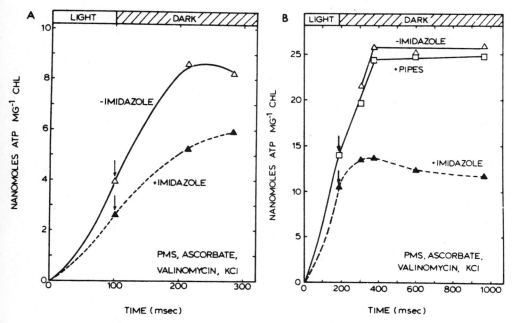

Fig. 2. The Effect of Imidazole on Photosynchronous and Postillumination Phosphorylation in the Presence of Valinomycin and K$^+$.

The effect of aniline on postillumination phosphorylation is shown in Fig. 4. There are several phases common to each curve, and the presence of aniline appears to extend each one. First is a period in which no ATP is synthesized and which is extended from 56 ms to 127 ms by aniline. Second is a nonlinear phase in which ATP yields continuously increase and which ends by approximately 1 nmole of ATP mg^{-1} of chlorophyll. This phase is also extended by aniline. The third phase is linear and represents the period during which the bulk of the ATP is made. The presence of aniline decreases the slope of this linear phase but also increases the extrapolated X-intercept from 177 ms to 267 ms. The fourth phase is a plateau which is lowered by aniline as can be seen in Fig. 5. If these effects of aniline are due to its ability to act as a buffer within the thylakoid lumen, then the effects should be exaggerated by an ensuing increase in the internal aqueous space, because although the aniline concentration would be the same, the total amount within the lumen would be increased. Fig. 5 shows the development of the capacity for postillumination phosphorylation when the sorbitol concentration in the medium is lowered from 0.2 to 0.02 M to bring about the volume change. As predicted, aniline lengthens the first phase (no ATP synthesis) from 127 ms to 248 ms, and increases the extrapolated X-intercept of the linear phase from 314 ms to 479 ms. Aniline appears to affect the initiation of postillumination ATP synthesis but not the initiation of photosynchronous ATP synthesis, and the weak amine buffers imidazole and pyridine affect these two types of ATP synthesis in similar manner (Horner and Moudrianakis, 1986).

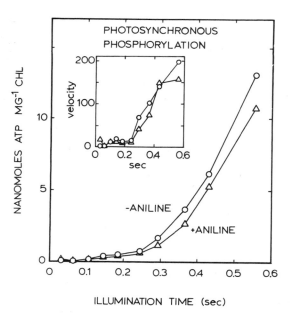

Fig. 3. The Effect of 13 mM Aniline on the Initiation of Photosynchronous Phosphorylation at pH 7.2 in the Presence of Valinomycin, K$^+$, and 0.2 M Sorbitol.

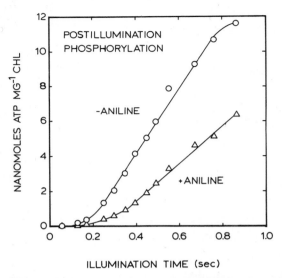

Fig. 4. The Effect of 13 mM Aniline on the Initiation of Postillumination Phosphorylation at pH 7.2 in the Presence of Valinomycin, K^+, and 0.2 M Sorbitol.

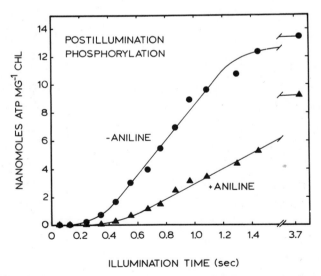

Fig. 5. The Effect of Aniline on the Initiation of Postillumination Phosphorylation at pH 7.2 in the Presence of Valinomycin, K^+, and 0.02 M Sorbitol.

These amines have been shown under some conditions to uncouple phosphorylation from electron transport in steady state experiments (Hind and Whittingham, 1963; Good, 1960), and the objection may be raised that a similar uncoupling here may be responsible for our results. However, if uncoupling is responsible for the delay observed in the initiation of postillumination ATP synthesis, it should also be seen in photosynchronous ATP synthesis, but it is not. The behavior of the system during photosynchronous phosphorylation can be considered as an internal control for postillumination phosphorylation.

Since stacks of thylakoids as are usually isolated from chloroplasts possess an extremely convoluted inner aqueous space, we determined to test for a buffer-induced delay of photosynchronous and postillumination phosphorylation in a defined population of subchloroplast particles, consisting of simple unilamellar vesicles capable of both photophosphorylation and postillumination phosphorylation (McCarty, 1971). Furthermore, instead of using weak amine buffers, we incorporated the zwitterionic buffers, MES and MOPS, within the vesicles during the sonication step in their preparation. These buffers do not uncouple (Good et al., 1966), as amines have been reported to do. Glycine was used as a control non-buffering zwitterion, and 45 mM NaCl was included to make the conductivity equal to the MES and MOPS media. The internalization of these zwitterionic buffers was confirmed by: (a) titration of the TCA supernatant of subchloroplast particles first desalted on G-25 Sephadex to remove external buffer; (b) by stimulation of Xe postillumination phosphorylation; (c) and by stimulation of light-induced proton uptake (Table I). The last two assays were used previously to demonstrate that amine buffers were internalized by chloroplasts (Nelson et al., 1971; Avron, 1972).

The effect of internalized MES and glycine on postillumination ATP synthesis is shown in Fig. 6. The capacity for postillumination phosphorylation increased as illumination time increased for both types of particles, and as predicted by the chemiosmotic hypothesis, the internal buffer MES decreased the initial amount of postillumination ATP synthesis. At 267 ms, both types of subchloroplast particles reached 75% of their maximum rates of photosynchronous ATP synthesis, and yet the glycine + NaCl subchloroplast particles had nearly four times the capacity for postillumination synthesis that the MES particles had. A similar situation occurred at 493 ms. MES-loaded subchloroplast particles appeared to support near maximum rates of photosynchronous ATP synthesis while at the same time exhibiting a marked inhibition of postillumination phosphorylation. This is consistent with our previous observations using aniline, pyridine, and imidazole as internal buffers in chloroplasts. Thus, internal buffers appear to have large effects on initial postillumination phosphorylation while having small or no effects on initial photosynchronous phosphorylation.

Table I. Stimulation of Postillumination Phosphorylation (Xe) and Light–Induced Proton Uptake in Chloroplasts and Subchloroplast Particles.

Buffer	Buffer pKa	Postillumin. Phosphorylation (nmol ATP/mg chl)	% Stim.	Light–Induced Proton Uptake (neq H$^+$/mg chl)	% Stim.	Ref.
CHLOROPLASTS						
Control		65	0%	398	0%	(1,2)
Pyridine	5.2	353	443%	1480	272%	(1)
Aniline	4.7	377*	480%*	2300	478%	(2)
Imidazole	6.8	139*	114%*	950**	1340%**	(2)
SUBCHLOROPLAST PARTICLES						
Control		6.6	0%	49	0%	
MES	6.2	10.7	62%	90	84%	

Postillumination phosphorylation (Xe) was assayed with the light phase at pH 6.0 and the dark phase at pH 8.5. *Light phase pH was 6.5 with a control value of 69 nmol ATP/mg chl. Light–induced proton uptake was assayed at pH 6.5. **Assayed at pH 8.0 with a control value of 66 neq H$^+$/mg chl. (1) Nelson et al. (1971). (2) Avron, M. (1972)

DISCUSSION

There have been numerous other studies of the initiation of ATP synthesis using either single millisecond light flashes or a series of single turnover (microsecond) light flashes to drive the electron transport system. The electrochemical potential of the system is altered in some way, and its effect on the initiation of ATP synthesis is estimated by the number of flashes necessary to observe the first ATP. The techniques employed in these studies are quite sensitive and sophisticated; however, our results from real-time rapid flow-quench measurements have necessitated a reinterpretation of some of the earlier conclusions. We have found that ATP synthesis <u>during</u> illumination appears to be driven by an electrochemical potential <u>distinct</u> from that which drives ATP synthesis <u>after</u> illumination. All studies which employ rapid light flashes or trains of flashes lack the potential of discrimination between photosynchronous and postillumination ATP synthesis. As long as the illumination time and the total reaction time remain different, yields of ATP will be a mixture of the results of two different modes of photophosphorylation, and conclusions about the nature of the developing electrochemical potential will be ambiguous.

Real-time rapid flow-quench techniques have allowed us to resolve the distinct effects which internal buffers have on the early (millisecond) events of the two components of photosynthetic ATP formation, i.e. photosynchronous ATP synthesis and postillumination (capacitance) ATP synthesis. From this resolution derive three observations which bear on the current question of localized vs. delocalized energy coupling. First, initial photosynchronous phosphorylation is not delayed by permeant buffers. Second, the development of the potential for postillumination (capacitance) phosphorylation is delayed by permeant buffers. Third, the gradual increase in the rate of photosynchronous phosphorylation during the first 500 ms of illumination indicates that, in the absence of a membrane potential, the utilization of the electrochemical potential by the ATP synthetase requires some hundreds of milliseconds to reach a steady state level. The canonical chemiosmotic view (Mitchell, 1966) of an obligatorily delocalized (global) proton gradient is consistent with the second and third observations but not the first. On the other hand, the concept of strictly localized proton pools driving ATP synthesis is inconsistent with the second observation, but

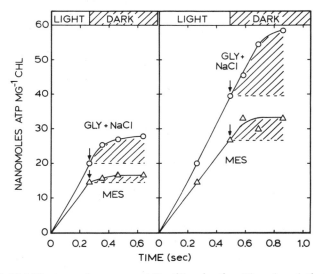

Fig. 6. Initial Photosynchronous and Postillumination Phosphorylation Using Subchloroplast Particles Made in the Presence of 0.1 M Glycine, 45 mM NaCl or 0.1 M MES at pH 8.0 in the Presence of Valinomycin and K^+. Under continuous flow, a suspension of subchloroplast particles mixed with phosphorylation medium as it was first illuminated, flowed through clear tubing under illumination for the time indicated, and at the arrow either mixed with perchloric acid or flowed into darkness within stainless steel tubing to allow ATP synthesis to continue for the indicated times before mixing with perchloric acid in darkness. The broken line marks the amount of ATP made in the light and the shaded area above that line represents the amount of postillumination ATP synthesis. Controls for steady state photophosphorylation had rates of 359 and 258 μmoles of ATP mg^{-1} of chlorophyll h^{-1} for glycine + NaCl and MES subchloroplast particles, respectively.

consistent with the first and conditionally so with the third.   If one combines the two concepts into one consisting of a single intramembrane proton pathway which branches into two readily equilibrating paths, one leading to the ATP synthetase and the other to the lumen (Dilley et al., 1981; Tandy et al., 1982; Dilley and Schreiber, 1984), this also will not account for the first observation.   Such a model with a branching point within the CF_o has been proposed by Dilley and coworkers (Fig. 1 of Dilley et al., 1981).   However, protons in the intramembrane path will not flow first to the ATP synthetase to perform chemical work if there exists another path (leading to the lumen) along which no work is performed.   A sizable transmembrane proton gradient would have to be established before the resulting back pressure would force protons to flow to the ATP synthetase.   Thus both photosynchronous and

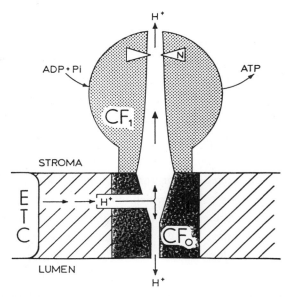

Fig. 7.   The CF_1-CF_o Complex Operating as a pKa Gate.   The gating mechanism depends upon a difference in the protonation potentials (pKa's) between the outward and inward paths which is illustrated as two channels of different widths. The higher pKa (greater tendency to protonate) of the CF_1 channel directs protons towards ATP synthesis without a dependence on the transmembrane proton gradient. As the chemical activity of the protons at the gate approaches steady state levels, the lower pKa part of the channel becomes protonated, ceases to be a barrier to flow, and "widens". This model allows the steady state transmembrane proton gradient to act as a capacitor to maintain constant rates of ATP synthesis at times when light intensity might fluctuate causing variations in rates of electron transfer and proton translocation. It also accounts for the observations that initial photosynchronous phosphorylation and postillumination phosphorylation appear to be driven by distinct parts of the electrochemical potential.

postillumination phosphorylation would be delayed by a internal buffer. To explain all three phenomena, a branching pathway is required, but at the branching point there must be a gate, limiting proton flow to the lumen, as shown in Fig. 7. One possible mechanism for the operational control of this gate could be that the two pathways would possess two different pKa's. If the chemical groups which formed the branch to the ATP synthetase had a pKa which is 2 pH units higher than the branch to the lumen, this difference in protonation potentials would counterbalance the absence of chemical work in the lumen pathway and make the ATP synthetase pathway competitive without completely stopping proton flow to the lumen. We have proposed the existence of such a "pKa gate" (Horner and Moudrianakis, 1986). The most efficient location for such this pKa gate would be within the CF_0-CF_1 domain, most likely within the CF_0. In the figure, the unshaded areas of the proton pathways are <u>not</u> meant to represent two <u>aqueous</u> <u>channels</u> of different diameters within the CF_0. On the contrary, they probably consist of two series of protonatable groups along one or more of the membrane-spanning helices of the CF_0 subunits. The different widths of the channels are simply meant to represent the relative proton conductance attributable to the relative protonation potentials of the groups making up the two pathways.

The idea of protons moving through a membrane is part of every model of energy transduction whether it involves global transmembrane proton partitioning (chemiosmosis) or intramembrane proton pools (localized coupling). The quinone pool is postulated to move protons, coupled to electron transport, through the membrane between photosystem II and I (Bouges-Bocquet, 1977). Even the movement of an unneutralized proton through a membrane is commonly observed. There are several membrane proteins which move protons through their hydrophobic domains within the membrane in the course of their catalytic activity. Bacterial rhodopsin, cytochrome oxidase and CF_0, are but three examples. Our concept of partitioning proton fluxes along intramembrane pathways shares principles with both coupling hypotheses. However, it is distinct from that of intramembrane proton pools and does not share with it the obvious thermodynamic drawbacks of accumulating mobile charged species within the hydrophobic span of the biological membrane. It also differs from chemiosmosis in that it considers the global transmembrane proton gradient as an end effect rather than a kinetic intermediate in the system, but it retains the vectorial proton transport of chemiosmosis although in a more complex path. However, once the basic principle of unneutralized proton movement through a hydrophobic environment is accepted, there is essentially no thermodynamic difference between protons transported across the membrane via some structure and protons moving transversely or otherwise through the membrane via another specialized structure, as is the case with the "pKa gate" concept of proton-flux partitioning presented here.

REFERENCES

Avron, M. (1972) in *Proceedings of the Second International Congress on Photosynthesis* (Forti, G., Avron, M. and Melandri, A., eds.) p. 861, Dr. W. Junk, by Publishers, The Hague.

Beard, W.A. and Dilley, R.A. (1986) FEBS Lett. 201, 57.

Bouges-Bocquet, B. (1977) Biochim. Biophys. Acta 462, 371.

Davenport, J.W. and McCarty, R.E. (1980) Biochim. Biophys. Acta 589, 353.

Dilley, R.A. (1972) Methods Enzymol. 24b, 68.

Dilley, R.A., Baker, G.M., Bhatnagar, D., Millner, P. and Lazlo, J. (1981) in *Energy Coupling in Photosynthesis* (Selman, B. and Selman-Reimer, S., eds.) p. 47, Elsevier/North Holland, New York.

Dilley, R.A. and Schreiber, U. (1984) J. Bioenerg. Biomemb. 16, 173.

Good, N.E. (1960) Biochim. Biophys. Acta 40, 502.

Good, N.E., Winget, G.D., Winter, W., Connolly, T.N., Izawa, S. and Singh, R.M.M. (1966) Biochemistry 5, 467.

Hind, G. and Whittingham, C.P. (1963) Biochim. Biophys. Acta 75, 194.

Horner, R.D. and Moudrianakis, E.N. (1983) J. Biol. Chem. 258, 11643.

Horner, R.D. and Moudrianakis, E.N. (1986) J. Biol. Chem. 261, 13408.

Jagendorf, A.T. and Uribe, E. (1966) Proc. Natl. Acad. Sci. USA. 55, 170.

Jagendorf, A.T. (1972) Methods Enzymol. 24b, 103.

Kell, D.B. (1979) Biochim. Biophys. Acta 549, 55.

McCarty, R.E. (1971) Methods Enzymol. 23, 302.

Mitchell, P. (1966) Biol. Rev. Cambridge Phil. Soc. 41, 455.

Nelson, N., Nelson, H., Naim, Y. and Neumann, J. (1971) Arch. Biochem. Biophys. 145, 263.

Ort, D.R., Dilley, R.A. and Good, N.E. (1976) Biochim. Biophys. Acta 449, 108.

Rottenberg, H. and Grunwald, T. (1972) Eur. J. Biochem. 25, 71.

Rottenberg, H. and Steiner-Mordoch, S. (1986) FEBS Lett. 202, 314.

Sigalat, C., Haraux, F., de Kouchkovsky, F., Hung, S.P.N. and de Kouchkovsky, Y. (1985) Biochim. Biophys. Acta 809, 403.

Tandy, N., Dilley, R.A., Hermondson, M.A. and Bhatnagar, D. (1982) J. Biol. Chem. 257, 4301.

Theg, S.M. and Junge, W. (1983) Biochim. Biophys. Acta. 723, 294.

Vinkler, C., Avron, M. and Boyer, P.D. (1980) J. Biol. Chem. 255, 2263.

Williams, R.J.P. (1978) FEBS Lett. 85, 9.

ATP SYNTHESIS DRIVEN BY INTRAMEMBRANAL PROTONS

Hagai Rottenberg, Todd P. Silverstein*, Ken Hashimoto**
and Sonia Steiner-Mordoch

Department of Pathology, Hahnemann University
Philadelphia, PA 19102

Recent evidence from several laboratories suggest the existence of a direct, intramembranal, proton pathway between the redox H^+-pumps and the ATPase H^+-pump in addition to the bulk to bulk pathway [reviewed in Rottenberg (1985), Ferguson (1985) and Westerhoff et al. (1984)]. We have suggested previously that collisions and dynamic aggregation of the mitochondrial inner membrane proteins enhance direct intramembranal proton transfer and thus energy conversion (Rottenberg 1978, 1985). We have shown previously that ambient temperature strongly affects the degree of coupling of oxidative phosphorylation in rat liver mitochondria. In high temperatures there is a significant reduction of both the respiratory control (Rottenberg, 1978) and the State 4 phosphate potential, ΔGp (Rottenberg et al. 1985). This reduction of respiratory control and phosphate potential at elevated temperatures is not due to reduction of $\Delta\mu_H$ which, in fact, increases slightly at elevated temperatures (Rottenberg et al. 1985). Thus, despite an increase in $\Delta\mu_H$, the efficiency of coupling, as reflected in the $\Delta Gp/\Delta\mu_H$ ratio, is decreased (Fig. 1, Rottenberg et al., 1985). Moreover, in liver mitochondria isolated from ethanol-fed rats (in which the concentration of the redox and the ATPase H^+-pumps are greatly reduced), the ratio $\Delta Gp/\Delta\mu_H$ is reduced at all temperatures even though $\Delta\mu_H$ is the same as in control rats. These data are compatible with the suggestion that efficient energy conversion depends on the frequency of collisions and the extent of dynamic aggregation of the inner membrane proteins. Mitochondrial membranes exhibit a sharp transition from a liquid crystalline to gel phase at a very

---

*Present address - Chemistry Department, Whitman College, Walla Walla, WA, 99632
**Present address - Wakunaga Pharmaceutical Co., Control Research Lab, Koda-Cho, Takata-Gun, Hiroshima 729-64, Japan

Molecular Structure, Function, and Assembly of the ATP Synthases
Edited by Sangkot Marzuki
Plenum Press, New York

low temperature (-11°C) and this phase transition is associated with a massive protein segregation as observed by electron microscopy (Hocheli and Hackenbrock, 1976). Additional, broad phase transition which is much weaker, is observed at higher temperatures (17°C - 33°C) by various techniques, including calorimetry (Blazk and Newman, 1980), lipid spin probes (Waring, et al., 1982) and membrane-enzymes arrhenius plots (Rottenberg et al., 1980). We have suggested that this secondary transition affects the extent of dynamic aggregation of the membrane proteins and thus the degree of coupling (Rottenberg, 1978). To study mitochondrial protein aggregation directly we have used the phosphorescent probe (Triplet probe)-erythrosin-isothiocyanate, which was covalently linked to proteins on the surface of inner membrane vesicles (submitochondrial particles). The decay of the phosphorescence anisotropy is biphasic (Silverstein and Rottenberg, 1987). An exponential phase decay at a rate which is proportional to the rotational diffusion of the proteins, while the residual anisotropy indicates the extent of aggregation of non-rotating proteins (Kawata et al., 1982, Coke et al., 1986).

We have studied the temperature dependence of this residual anisotropy in erythrosin-isothiocyanate labeled submitochondrial particles. The residual anisotropy show a broad transition at 15-35°C from a state of highly aggregated proteins (at low temperature) to a state of slightly aggregated proteins (at high temperature). To ascertain that the aggregation of the ATPase is similar to other membrane proteins we

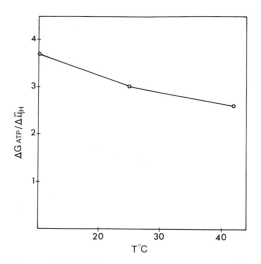

Fig. 1. The Efficiency of Oxidative Phosphorylation in Rat Liver Mitochondria as Function of Temperature. The ratio of the free energy of ATP hydrolysis at State 4 (ΔGp) and the proton electrochemical potential ($\Delta \tilde{\mu}_H$) was measured as function of temperature. ΔGp was estimated from a "null-point" titration of the rate of phosphorylation against the ratio [ATP]/[ADP][Pi]. $\Delta \tilde{\mu}_H$ was estimated by flow dialysis from the distribution of [^{14}C]DMO (ΔpH) and ^{86}Rb + valinomycin ($\Delta \psi$). From Rottenberg et al., 1985.

have labeled specifically the F_1 portion of the ATPase by removing the F_1 from the membranes (Linet et al., 1979) reacting it with erythrosin-isothiocyanate and reconstituting the labeled F_1 with urea-treated submitochondrial particles [which are stripped of the F_1 (Pedersen and Hullihen, 1979)]. Fig. 2 shows the temperature dependence of the residual anisotropy measured 0.8 msec after a short (5 μsec) flash in the reconstituted vesicles (Silverstein and Rottenberg, 1987).

At very low temperatures (below -11°C) the residual anisotropy is very high, indicating massive aggregation, as observed by electron microscopy (Hocheli and Hackenbrock, 1976). Increasing the temperature above the phase transition sharply reduces the residual anisotropy to an intermediate level. A further increase of temperature shows another broad transition at 15-35°C which leads to a low level of residual anisotropy. These transitions are completely reversible. However, when the temperature is increased further, above 50°C, protein denaturation occurs, which is also associated with protein aggregation. This process is irreversible and cannot be reversed by lowering the temperature (Fig. 2). Recently, we have also used another approach to study temperature effect on protein aggregation. We have first labeled

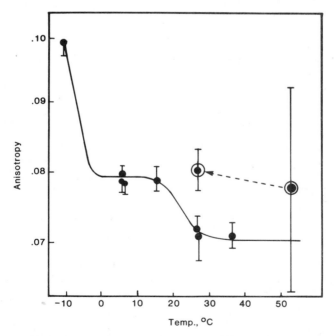

Fig. 2. ATPase Aggregation in Mitochondrial Inner Membrane Vesicles as Function of Temperature. The residual anisotropy of the phosphorescence of F_1 labeled with erythrosin-isothiocyanate and reconstituted with urea-treated submitochondrial particles was studied as function of temperature. Phosphorescence anisotropy was measured in anaerobic suspension on a Perkin-Elmer LS-5 instrument with a delay time of 0.8 msec and a gating time of 1.0 msec (Silverstein and Rottenberg, 1987).

197

the F_o portion of the ATPase with the fluorescent DCCD analog, NCD-4 (Pringle and Taber, 1985). We then used DPH, a lipid phase probe, to study fluorescence energy transfer between these two dyes in the inner membranes. Since the efficiency of transfer strongly depends on the distance between donor and acceptor, aggregation of the ATPase should reduce the efficiency. Indeed, the efficiency of energy transfer increased with temperature, with a transition overlaping the broad transition (15°-35°C) shown in Fig. 2 (results not shown). These results, taken together with the temperature dependence of the efficiency of oxidative phosphorylation, strongly suggest that dynamic aggregation enhances energy conversion in oxidative phosphorylation and thus supports the notion that direct intramembranal proton transfer contributes significantly to proton driven ATP synthesis.

Another approach to the study of intramembranal proton transfer is to utilize decouplers, i.e., uncoupling agents that do not collapse $\Delta\mu_H$. We have previously shown that general anesthetic stimulates the rates of state 4 respiration and ATPase and inhibits phosphorylation without a significant effect on $\Delta\mu_H$ (Rottenberg, 1983). We have screened a large number of reagents and identified several classes of decouplers (Hashimoto and Rottenberg, unpublished). One class of decouplers are detergents, particularly Triton X-100 which uncouple oxidative phosphorylation at an

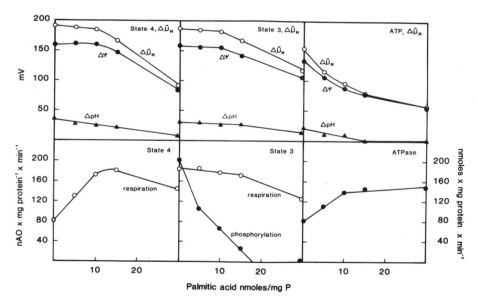

Fig. 3. Uncoupling of Oxidative Phosphorylation in Rat Liver Mitochondria by Palmitic Acid. $\Delta\mu_H$ was estimated from the distribution of [^{14}C]DMO (ΔpH) and ^{86}Rb + valinomycin ($\Delta\psi$). ATP synthesis was measured from ^{32}P incorporation into ATP and ATP hydrolysis from the disappearance of ATP-^{32}P. Respiration was measured by oxygen electrode. All measurements were at 37°C. From Rottenberg and Hashimoto, 1986.

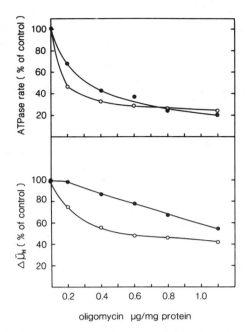

Fig. 4. Effect of Palmitate on the Oligomycin Sensitivity of Rat Liver Mitochondria. The rate of ATPase and $\Delta\mu_H$ (generated by ATP hydrolysis) were measured as function of oligomycin concentration with palmitic acid (10 nmoles/mg protein, o) and without (●). Assays as in Fig. 3. From Rottenberg and Hashimoto, 1986.

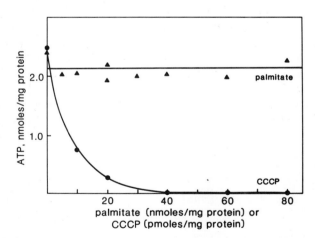

Fig. 5. The Effect of Palmitate and CCCP on ATP Synthesis Induced by Artificial $\Delta\mu_H$ in Submitochondrial Particles from Rat Liver. Artificial $\Delta\mu_H$ was generated by imposition of a transient pH gradient and a potassium diffusion potential. Phosphorylation was measured as in Fig. 3. From Rottenberg and Steiner-Mordoch, 1986.

extremely low concentration range without effect on $\Delta\mu_H$ (Hashimoto and Rottenberg, unpublished). Preliminary results suggest that Triton X-100 decouple specific intramembranal proton transfer by dispersing the protein aggregates (Silverstein and Rottenberg, 1987). Another class of decouplers, which appear to work by an entirely different mechanism, are free fatty acids (FFA). FFA also uncouple oxidative phosphorylation without effect on $\Delta\mu_H$ (Rottenberg and Hashimoto, 1986). Fig. 3 shows the characteristics of this uncoupling process. State 4 respiration is stimulated to the level of State 3 respiration without significant reduction of $\Delta\mu_H$, phosphorylation, at State 3 is inhibited without significant effect on $\Delta\mu_H$. However, when ATP is the substrate a stimulation of ATPase is associated with a reduction of $\Delta\mu_H$. The efficiency of decoupling by FFA increases at high temperature, suggesting that proton aggregation hinder the access of FFA to the proton pumps. Since FFA reduces the $\Delta\mu_H$ generated by ATP, but has no effect on electron transport driven $\Delta\mu_H$, it is suggested that there is a direct interaction between FFA and the ATPase. This conclusion is also supported by the finding that FFA enhance the effect of oligomycin on the ATPase activity and the $\Delta\mu_H$ generation by ATP (Fig. 4). However, FFA are not intrinsic uncouplers of the H^+-ATPase since they have no effect on ATP synthesis driven by artificially imposed $\Delta\mu_H$ (Fig. 5, Rottenberg and Steiner-Mordoch, 1986). Thus, FFA appear only to uncouple phosphorylation driven by intramembranal proton transfer and are without effect on phosphorylation driven by the bulk $\Delta\mu_H$. Similar effects of FFA on photophosphorylation in chloroplasts were obtained recently (Pick et. al., submitted). We have also observed recently that FFA enhance the interactions between DCCD and the F_o-portion of the ATPase. Thus, the data obtained so far suggest that FFA decouple by specific interaction with the H^+-ATPase which prevent the H^+-ATPase from utilizing intramembranal protons derived from redox pumps.

We have previously suggested that the F_o portion of the ATPase and more specifically the DCCD-binding protein (subunit c) serves as a proton reservoir in the ATPase complex (Rottenberg, 1985). This reservoir would enable the complex to accept proton from neighboring redox pumps directly, and to feed them into the proton channel (wire) where the proton transfer is linked to ATP synthesis. This suggestion can now be translated into a more specific working hypothesis for the proton conducting segments of the F_o portion of the ATPase. This working model (Fig. 6) incorporate recent suggestions that indicate that the carboxy terminal segment of subunit a (subunit 6) constitutes the proton channel (more appropriately described as proton conducting wire) (Fox et al. 1986; Cain and Simoni, 1986), and that one of the DCCD-binding groups serve to connect and possibly gate two portions of the wire (Fox et al., 1986). Fig. 6A shows a sectional view of the proton conducting segment of the F_o. The transport of proton across the membrane proceed through a proton conducting pathway ("wire") on the carboxy terminal segment of subunit 6 (*a*). There are two segments leading from both sides of the membrane to the membrane core. For a proton to pass from one segment to the other it is necessary for it to bind to the carboxylic residue glu-61 (asp-61) of the DCCD-binding protein (subunit *c*). It is assumed that this step is coupled to a conformational transition and energy conversion. The DCCD-binding proteins can freely exchange protons with all the other copies of this peptide by proton jumping laterally in the core of the membranes and thus serve as proton reservoir. This reservoir is accessible also to similar groups on redox pumps and the transfer of protons laterally in the core of the membranes between the redox pumps and the ATPase is enhanced by protein collision and dynamic aggregation.

Fig. 6B shows a top view of the proton conducting segments of subunit 6 (*a*) and the DCCD-binding protein demonstrating the feasibility of exchange of proton between residue 61 of the DCCD-binding proteins. Helices a_4, a_5 of subunit *a* contain the transmembrane H^+-pathway and helices c_1 of subunits *c* contain the proton binding residue 61 and form the proton pool. The model can accommodate any number of subunits *c*. Our model differs somewhat from the model of Cox et al. (1986) in the placement of the other copies of subunit *c*, which we suggest to form a pocket attached to the channel and which serve as (a) proton reservoir and (b) proton donor (or acceptor) to other pumps. This model is compatible with the fact that the number of

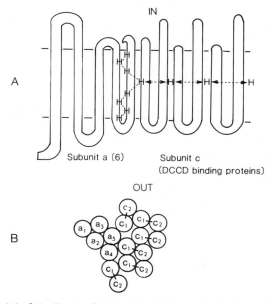

Fig. 6. A Model of the Proton Conducting Pathway, Including a Proton Pool in the F_0. 6A shows a section through the membrane of subunit 6(a) and adjacent DCCD-binding proteins (subunit c). 6B shows a top view from the membrane surface of the same segment of F_0. See text for more detail.

copies of subunit c in F_o appear to vary greatly between species and possibly between various physiological states without a great effect on function (Hoppe and Sebald, 1984) and that the DCCD-binding protein appear to be freely accessible to lipophilic reagents (Hoppe et al. 1984). Moreover, this model allow proton transport to occur both across the membrane (through the subunit a wire) and laterally in the membrane core (through the pool of subunit c), as postulated in the parallel coupling model (Rottenberg, 1985). In addition to accommodating the suggestion of intramembrane H^+ transfer, it could explain the fact that some F_o mutants, while deficient in transmembrane proton transfer through the channel (wire) show normal oxidative phosphorylation (Cox et al. 1983). The uncoupling effect of FFA is suggested to be the result of specific interactions with the DCCD-binding proteins which cause a proton leak out of the intramembranal proton "pool". Studies, now in progress, on the interactions between fatty acids and the DCCD-binding protein should lead to better understanding of the function of these subunits in proton storage and conduction during oxidative phosphorylation.

ACKNOWLEDGMENTS

Supported by PHS grants GM-28173 and AA07238.

REFERENCES

Blazk, J.F. and Newman, J.L. (1980) Biochim. Biophys. Acta 600, 1007-1011.

Cain, D.B. and Simoni, R.D. (1986) J. Biol. Chem. 261, 10043-10050.

Cox, G.B., Jans, D.A., Gibson, F., Langman, L., Senior, A.E. and Fimmel, A.L. (1983) Biochem J. 216, 143-150.

Coke, M., Restall, C.J., Kemp, C.M. and Chapman, D. (1986) Biochemistry 25, 513-518.

Cox, G.B., Fimmel, A.L., Gibson, F. and Hatch, L (1986) Biochim. Biophys. Acta 849, 62-69.

Ferguson, S.J. (1985) Biochim. Biophys. Acta 811, 47-95.

Hochli, M. and Hackenbrock, G.R. (1976) Proc. Natl. Acad. Sci. USA 76, 1236-1240.

Hoppe, J. and Sebald, W. (1984) Biochim. Biophys. Acta 768, 1-27.

Hoppe, J., Brunner, J. and Jorgensen B.B. (1984) Biochemistry 23, 5610-5616.

Kawata, S., Lehner, C., Muller, M., Cherry, R.J. (1982) J. Biol. Chem. 257, 6470-6476.

Linnet, P.E., Mitchell, A.D., Portis, M.D. and Beechey, R.B. (1979) Methods Enzymol. 55, 337-343.

Pedersen, P.L. and Hullihen, J. (1979) Methods Enzymol. 55, 736-741.

Pringle, M.J. and Taber, M. (1985) Biochemistry 24, 7366-7371.

Rottenberg, H. (1978) FEBS Lett. 94, 295-297.

Rottenberg, H. (1983) Proc. Natl. Acad. Sci. USA 80, 3313-3317.

Rottenberg, H. (1985) Modern Cell Biol. 4, 47-83.

Rottenberg, H. and Hashimoto, K. (1986) Biochemistry 25, 1747-1755.

Rottenberg, H. and Steiner-Mordoch. S. (1986) FEBS Lett. 202, 314-318.

Rottenberg, H., Robertson, D.E. and Rubin, E. (1980) Lab. Inv. 42, 318-326.

Rottenberg, H., Robertson, D.E. and Rubin, E. (1985) Biochim. Biophys. Acta. 809, 1-10.

Silverstein, T. and Rottenberg, H. (1987) Biophysical J. 489a.

Waring, A.J., Rottenberg, H., Ohnishi, T. and Rubin, E. (1982) Arch. Biochem. Biophys. 216, 51-61.

Westerhoff, H.V., Melandri, B.A., Venturoli, G., Azzone, G.F. and Kell, D.B. (1984) Biochim. Biophys. Acta 768, 257.

ATP SYNTHASE REACTION MECHANISMS

CATALYSIS BY ISOMERIC FORMS OF COVALENTLY LABELED F_1-ATPase

Jui H. Wang, Joe C. Wu, Vijay Joshi, Hua Chuan and Jennifer Cesana

Bioenergetics Laboratory, Acheson Hall, State University
of New York, Buffalo, NY 14214 U.S.A.

INTRODUCTION

When bovine heart mitochondrial F_1-ATPase was labeled with low concentrations of NBD-Cl* in the dark in the presence of ATP at pH 7, the NBD-label was almost exclusively attached to a specific tyrosine residue until about one label is covalently attached per F_1 molecule (Ferguson et al., 1975a). Cleavage of the labeled protein, isolation and sequencing of the labeled polypeptide and comparison with the known amino acid sequence of the β subunit of the F_1 showed that the NBD-label is initially attached to Tyr-β311 (Runswick and Walker, 1983; Andrews et al., 1984a; Sutton and Ferguson, 1985b). Several challenging questions concerning these observations remain unanswered. First, why does the NBD-label inhibit the ATPase? Is it because Tyr-β311 is at the catalytic site so that the label interferes directly with the catalytic mechanism, or because Tyr-β311 is not at the catalytic site but its labeling triggers a long-range protein conformation change which inactivates the enzyme? Second, inasmuch as there are three intrinsically identical β subunits in each F_1 molecule, why does the labeling of one β subunit per F_1 inactivates the whole enzyme?

---

*The abbreviations used are: AC, N-acetyl-L-cysteine; AMEDA, P^1-(5'-adenosyl)-P^2-N-(2-mercaptoethyl)diphosphoramidate; ASU, submitochondrial particles prepared from bovine heart mitochondria by sonication at pH 9 followed by steps involving urea treatment; DCCD, N,N'-dicyclohexyl-carbodiimide; DTT, DL-dithiothreitol; FCCP, carbonylcyanide p-(trifluoromethoxy)phenyl-hydrazone; n, molar ratio of label to F_1; NBD-Cl, 7-chloro-4-nitro-2,1,3-benzoxadiazole; O-NBD-F_1, F_1 labeled with NBD-Cl at its Tyr-β311 residue; r, ratio of the specific activity of the labeled enzyme to that of the unlabeled control; SMP, submitochondrial particles.

Molecular Structure, Function, and Assembly of the ATP Synthases
Edited by Sangkot Marzuki
Plenum Press, New York

LOCATION OF THE NBD-LABEL RELATIVE TO THE BOUND SUBSTRATE

In order to answer the first question, the compound AMEDA with its structure shown in Fig. 1 was synthesized and studied (Wu et al., 1987). It is well-known that the NBD-label on Tyr-β311 can be rapidly removed by sulfhydryl compounds such as dithiothreitol (DTT) or N-acetyl-L-cysteine (AC). If Tyr-β311 is at the hydrolytic site, we would expect the ATP analogue to be bound very close to the O-[^{14}C]NBD-label so that its sulfhydryl group could react *in situ* with the latter to form AMEDA-[^{14}C]NBD and regenerate the active F_1-ATPase as illustrated.

It was indeed found in a series of experiments that AMEDA reactivated O-NBD-F_1, and the reactivation was effectively prevented by either ADP or ATP. A kinetic analysis of the data showed that AMEDA first binds to O-NBD-F_1 with k_d = 15 μM, then reacts *in situ* with the label to produce AMEDA-NBD and the reactivated enzyme. The product AMEDA-NBD was isolated, identified and shown to bind even more tightly to F_1 with k_d = 2 μM. These observations and the structural similarity between AMEDA and ATP may be regarded as compelling evidence for the presence of the NBD-labeled Tyr-β311 near the phosphate groups of ATP bound at the hydrolytic site of F_1-ATPase, because it would be inconceivable for an AMEDA molecule bound at the hydrolytic site to react with a far away NBD-label to form the product AMEDA-NBD and then to be bound again to the site with even higher affinity (Wu et al., 1987). Since the NBD-label on Tyr-β311 can be transferred spontaneously to Lys-β162 at pH $\geq$ 9 (Ferguson et al., 1975b; Andrew et al., 1984b), we may conclude that Lys-β162 is also at the hydrolytic site of F_1-ATPase.

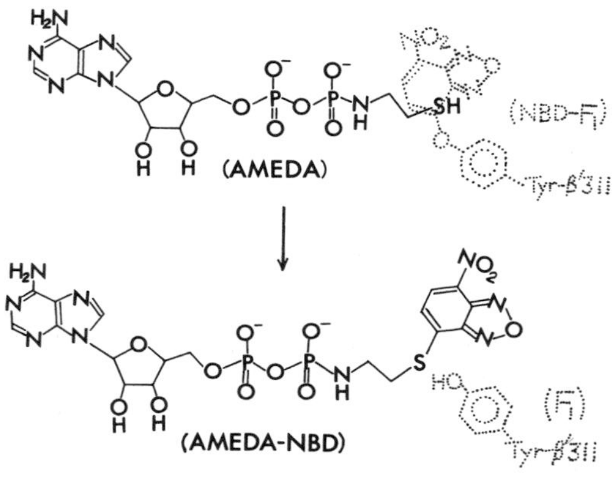

Fig. 1. *In Situ* Removal of NBD-Label by Bound AMEDA in F_1-ATPase.

ARE THE β SUBUNITS FUNCTIONALLY EQUIVALENT?

The observation that a single NBD-label attached to one Tyr-β311 is sufficient to inhibit completely the ATPase activity of F_1 has been interpreted in two ways. According to the alternating sites model (Gresser et al., 1982), the rapid hydrolysis of an ATP molecule bound to an active site requires a protein conformation at the site coupled to ligand change at two equivalent sites which alternates with the first. Labeling the Tyr-β311 at any one of the three equivalent sites would block one of the necessary consecutive steps in the catalytic cycle and thereby stop the steady-state hydrolysis of ATP.

According to the model with one hydrolytic site and two regulatory sites (Wang, 1984), the rapid hydrolysis of an ATP molecule bound to the hydrolytic site in β' also requires a protein conformation which is stabilized by the adenine nucleotides bound at the regulatory β'' sites. But unlike in the alternating sites model, the nucleotides bound at the β' and β'' sites in this model are not required to turn over at the same rate during the steady-state hydrolysis of ATP. Direct labeling of F_1 by NBD-Cl takes place predominantly at the β' site because of the enhanced reactivity of its Tyr-β311 phenolic group. Therefore, according to this model the F_1 would also be inhibited completely by NBD-Cl at a label to F_1 ratio of n = 1. Although the three β subunits of F_1 are intrinsically identical, their interaction with the smaller subunits γδε could make them differ functionally. With the regulatory sites on β'' occupied by adenine nucleotides, interaction between the subunits could make the hydrolytic efficiency of β' (multi-site catalysis) several orders of magnitude higher than when they are empty (uni-site catalysis).

In order to test these two proposed models for F_1, a procedure was developed to induce rearrangement of the subunits in F_1. It was found that when unlabeled F_1 was treated with 3 M LiCl for 2-4 min at 0°C and subsequently separated from LiCl in the presence of 5 mM ATP at 25°C by centrifugal gel-filtration (Wang, 1985; Wang et al., 1986), most of the ATPase activity of the F_1 could be retained. It was hoped that this brief LiCl-treatment could cause the subunits to rearrange and change their partners of close interactions. For O-NBD-F_1 with n = 1, the rearrangement could be as illustrated in Fig. 2.

Since the ATPase activity of unlabeled F_1 is also decreased by the LiCl-treatment because of partial denaturation, the observed changes in ATPase activity of the labeled enzyme is best expressed in terms of relative activity defined by

$$r = \frac{\text{specific activity of labeled } F_1}{\text{specific activity of control unlabeled } F_1}$$

The control unlabeled F_1 could be a sample of unlabeled F_1 which had been subjected to the same treatment. But a better control would be an aliquot of the labeled F_1 with all of its label removed by DTT at the last step. According to the alternating sites model, r should not be affected by the LiCl-treatment, since all β subunits are assumed to be equivalent. According to the other model, the LiCl-treatment could cause the labeled subunit in the hydrolytic β' position to switch place with an unlabeled subunit in the regulatory β'' position and hence could increase the value of r. Such experiments have been performed many times with highly reproducible results. A good example is given in Table I. Analyses of 11 labeled samples after the LiCl-treatment show that the number of label per F_1 was still n = 1.02 ± 0.04. Table I shows that the LiCl-treatment can raise the value of r from 0.03 to 0.62 without removing the covalent label. These results contradict the alternating sites model, but are consistent with the model with one hydrolytic and two regulatory sites.

GEOMETRIC ISOMERS OF COVALENTLY LABELED F_1-ATPase

The NBD-label on Tyr-β'311 can be removed at pH 7.0 in the dark with N-acetyl-L-cysteine in a well-controlled way. For example, when $(O-\beta'-NBD)_{1.0}F_1$ was incubated with 20 μM of N-acetyl-L-cysteine at room temperature, the value of r increases as the number n of label per F_1 decreases. The plot of r versus n is linear with a slope equal to -1 (see Fig. 3), which shows that the removal of each NBD-label completely reactivates an F_1 molecule. Treatment of $(O-\beta'-NBD)_{1.0}F_1$ with LiCl increased its r value from less than 0.1 to about 0.6 without changing n, presumably because some of the labeled subunit in β' position had been rearranged to the regulatory β'' position and some unlabeled subunit in β'' position had been rearranged

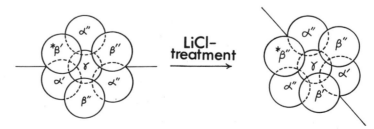

Fig. 2. Possible Subunit Rearrangement in O-NBD-F_1-ATPase Induced by LiCl-Treatment. The asterisk represents a Tyr-β311-[^{14}C]NBD-label. The straight line represents the C_2-axis of approximate molecular symmetry observed at low resolution (Amzel et al., 1982).

Table I. Effect of LiCl-treatment on the ATPase Activity of $(Tyr-\beta'311-[^{14}C]NBD)_{1.02}F_1$.

	S.A. (μmol ATP mg^{-1}min^{-1})		$r = \dfrac{\text{S.A. of NBD-F}_1}{\text{S.A. of F}_1}$
	-DTT	+DTT	
No LiCl-treatment	1.48	43.3	0.033
Scrambling with 3 M LiCl at 0°C	16.2	25.4	0.62

to the hydrolytic β' position as illustrated in Fig. 2. Treatment of the rearranged enzyme O-β', β''-NBD-F$_1$ with 20 μM N-acetyl-L-cysteine removed the NBD-label at the same rate as before until all labels on β' subunits were removed. Further removal of NBD-label from β'' subunits progressed at about 10-fold slower rate and no longer affected the value of r.

If the reaction mixture containing the rearranged sample was gel-filtered as soon as all NBD-labels on β' subunits had been removed, a labeled enzyme containing ~ 0.6 label/F$_1$ was obtained with r = 1 and presumably the formula (O-β''-NBD)$_{0.6}$F$_1$. On the other hand, if the reaction mixture containing the unscrambled sample was gel-filtered as soon as n had decreased to 0.5 to 0.6, a labeled enzyme was obtained with r = 0.5 to 0.6 and presumably the formula (O-β'-NBD)$_{0.6}$F$_1$. In this way, two geometric isomers of covalently labeled F$_1$ were obtained with contrasting biochemical properties. At pH 9 in the dark, the NBD-label in O-β'-NBD-F$_1$ was found to transfer

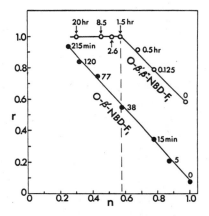

Fig. 3. Selective Removal of O-NBD-Label from O-NBD-F$_1$ by N-Acetyl-L-Cysteine in the Dark at 25°C (Wang et al., 1986).

spontaneously from Tyr-β311 to Lys-β162 with a half-life of ~2 hr; whereas, the NBD-label in O-β"-NBD-F$_1$ was found not to transfer at all under the same conditions. Both the horizontal portion of the biphasic linear plot in Fig. 3 and the observed slow O → N transfer rate for O-β"-[^{14}C]NBD-F$_1$ show that the latter was a pure isomer with contrasting property, not a mixture of labeled and unlabeled forms of the enzyme. Cleavage of O-β'-[^{14}C]NBD-F$_1$ or O-β"-[^{14}C]NBD-F$_1$ with pepsin, separation of the radioactive peptide and subsequent determination of its amino acid sequence showed that the NBD-label was covalently attached to Tyr-β311 in both geometric isomers (Wang et al., 1986).

One could speculate that although the NBD-label was still covalently attached to Tyr-β311, perhaps the treatment with LiCl had changed it to a non-interfering conformation so that the ATPase was partially reactivated. If this were the case, the Tyr-β311 residue which had already been labeled with a non-interfering NBD-group could not be labeled again. On the other hand, if the LiCl-treatment had indeed caused a labeled subunit in the β' position to switch place with an unlabeled subunit in the β" position, the new unlabeled subunit in β' position should again be susceptible to labeling by NBD-Cl. Fig. 4 shows that the rearranged enzyme O-β', β"-NBD-F$_1$ can be labeled again by NBD-Cl in the predictable way.

CATALYTIC HYDROLYSIS AND SYNTHESIS OF ATP BY Tyr-β'311-NBD-F$_1$-ASU

It is of interest to find out whether the catalytic hydrolysis and synthesis of ATP take place in opposite directions along the same reaction path. One way to test this possibility is to compare catalytic efficiency for ATP hydrolysis with that for ATP synthesis by submitochondrial particles containing stereospecifically labeled F$_1$. In order to avoid unnecessary complications, the soluble F$_1$-ATPase was first labeled with [^{14}C]NBD-Cl and then used to reconstitute with ASU to form various types of chemically modified submitochondrial particles (Steinmeier and Wang, 1979).

Oxidative phosphorylation can be measured only with membranes in the energized state, but ATP hydrolysis is often measured with uncoupled SMP in the deenergized state. Thus we should first examine the effect of energization on the ATPase activity, before a valid comparison could be made between the catalytic efficiency for ATP hydrolysis and that for ATP synthesis by chemically modified submitochondrial particles. For this purpose, the specifically labeled enzyme (O-β'-NBD)$_n$F$_1$ with n = 1.02 and r = 0.03 was used to reconstitute with ASU. The rate of catalytic ATP hydrolysis by the reconstituted SMP was measured both directly in uncoupled SMP in the presence of 12.5 μM FCCP and indirectly in coupled SMP through the coupled reduction of NAD$^+$ by succinate. The results summarized in Table II show that the ratio r of the specific activity of the labeled enzyme complex to that of the unlabeled control is essentially independent of the state of energization, and is determined only by the state of specific labeling of F_1.

The observed r values for catalytic ATP hydrolysis and oxidative phosphorylation by a similar pair of labeled and unlabeled SMP are summarized in Table III. The r value for oxidative phosphorylation is clearly much higher than that for ATP hydrolysis. Since the respiration rates of F_1-ASU and O-β'-NBD-F_1-ASU are essentially the same (Steinmeier and Wang, 1979), the data suggest that oxidative phosphorylation may involve more active sites of F_1 than catalytic ATP hydrolysis. In other words, the two catalytic processes may not have precisely the same reaction path as it is assumed in the alternating sites model (Gresser et al., 1982). For the present system F_oF_1-catalyzed ATP synthesis seems to have more reaction paths than ATP hydrolysis.

IS INTERNAL ROTATION NECESSARY FOR OXIDATIVE PHOSPHORYLATION?

As an alternative interpretation, we could assume that only β' catalyzes both the hydrolysis and synthesis of ATP, but that during oxidative phosphorylation the β subunits were scrambled by the proton flux so that both the labeled and unlabeled β subunits had their turn in the catalytic position (β') as well as in the regulatory positions (β"). If the functional differentiation of β' and β" is due to a closer interaction of β' with the smaller subunits, such speculation would be equivalent to assuming that the proton flux causes internal rotation of $\alpha_3\beta_3$ moiety relative to the $\gamma\delta\epsilon$ moiety of F_1 during oxidative phosphorylation. Accordingly, the r value of the

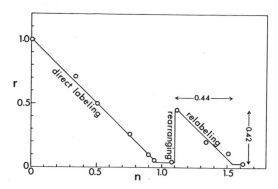

Fig. 4. Changes in n and r Due to Direct Labeling of F_1-ATPase by NBD-Cl, Li-Cl-Induced Rearrangement, and Subsequent Relabeling (Wang et al., 1986).

213

Table II. ATP Hydrolysis and Reverse Electron Transport by Chemically Modified SMP at 30°C.

Measurement	F_1-ASU	O-β'NBD-F_1-ASU	r
ATPase (- oligomycin, + FCCP) (μmol ATP min^{-1}mg^{-1})	5.15 ± 0.10	0.46 ± 0.004	0.090
Reverse electron transport (+ 0.25 μg oligomycin mg^{-1}) (nmol NADH min^{-1}mg^{-1})	29.4	2.78	0.095

unrearranged samples for oxidative phosphorylation is expected to be higher than that for ATP hydrolysis.

In order to check this possibility, the reconstituted SMP samples were preincubated under simulated oxidative phosphorylation conditions and subsequently assayed for ATPase activity (Wang et al., 1987). Table IV shows that the observed value of r = 0.23 for catalytic ATP hydrolysis by submitochondrial particles which had been preincubated for 12 min under simulated oxidative phosphorylation conditions is higher than the value r = 0.16 for the unpreincubated samples only by a factor of 1.4. This increase in r by preincubation is about the same as that observed in previous experiments with DCCD-labeled SMP (Soong and Wang, 1984). On the other hand, the value of r = 0.43 for oxidative phosphorylation (Table III) is higher than that for ATP hydrolysis by a factor of 3.6. Had the β subunits indeed switched roles in each catalytic cycle, we would expect the r for ATP hydrolysis by the preincubated samples also to increase by a factor of 3.6, because the F_0F_1 complex turned over 40000 times during the 12-min preincubation period. These results show that β' and β" subunits did not switch roles, and hence internal rotation of $\alpha_3\beta_3$ relative to $\gamma\delta\epsilon$ in F_1 did not take place during oxidative phosphorylation.

CATALYTIC HYDROLYSIS AND SYNTHESIS OF ATP BY Tyr-β"311-NBD-F_1-ASU

A more decisive test of the possibility that more active sites on F_1 may be involved in oxidative phosphorylation than in catalytic ATP hydrolysis could be conducted if F_1-ATPase with [^{14}C]NBD-labeled Tyr-311 in β" subunits and unlabeled β subunit could be used to reconstitute the submitochondrial particles. Because of their low reactivity, the Tyr-β311 in β" subunits of F_1 cannot be labeled directly by [^{14}C]NBD-Cl with satisfactory specificity. But by using three cycles of the label-

Table III.   ATP Hydrolysis and Oxidative Phosphorylation by Chemically Modified SMP at 30°C

Measurement	F_1-ASU	O-β'-NBD-F_1-ASU	r
ATPase (μmol ATP min^{-1}mg^{-1})	3.51 ± 0.09	0.43 ± 0.08	0.12
Oxidative Phosphorylation (nmol ATP min^{-1}mg^{-1})	111 ± 5	48 ± 2	0.43

rearrange-relabel process, the labeled enzyme (Tyr-β"311-[^{14}C]NBD)$_{1.3}$-F_1 was finally prepared and used to reconstitute with ASU particles (Wong et al., 1987). The observed relative catalytic efficiencies for the hydrolysis and synthesis of ATP by these reconstituted SMP are summarized in Table V.

The r-values in Table V show that the NBD-label attached to Tyr-β"311 inhibits oxidative phosphorylation but does not inhibit catalytic ATP hydrolysis in the reconstituted SMP. This observation further shows that although only the β' subunit in F_1 catalyzes ATP hydrolysis directly, both β' and β" catalyze oxidative phosphorylation.

It is not clear how all three β subunits of F_1 in the F_oF_1 complex can catalyze ATP synthesis, whereas only the β' subunit catalyzes efficient ATP hydrolysis directly. One possibility is that the high ATPase activity of isolated F_1 or reconstituted F_oF_1 could be an artifact of the preparative procedure. Interaction between the β' and β"

Table IV.   Effect of Preincubation Under Simulated Oxidative Phosphorylation Conditions of the ATPase Activity of SMP at 30°C

Measurement	F_1-ASU	O-β'-NBD-F_1-ASU	r
ATPase without preincubation (μmol ATP min^{-1}mg^{-1})	4.02	0.622 ± 0.009	0.16
ATPase after 12-min preincubation under ox. phos. conditions (μmol ATP min^{-1}mg^{-1})	4.08	0.917 ± 0.023	0.23

Table V. Catalytic Activities of SMP Reconstituted from $(\text{Tyr-}\beta''311\text{-}[^{14}\text{C}]\text{NBD})_{1.3}\text{F}_1$ and ASU at 30°C.

Measurement	Control F_1-ASU	$(\text{O-}\beta''\text{-}[^{14}\text{C}]\text{NBD})_{1.3}\text{F}_1$-ASU
ATP hydrolysis (μmol ATP min^{-1}mg^{-1})	1.88	1.84 r = 0.98
Oxidative phosphorylation (nmol ATP min^{-1}mg^{-1})	269	160 r = 0.61

subunits in the isolated F_1 or reconstituted F_0F_1 could provide only the catalytic site on β' with a unique mechanism for directly catalyzing the rapid formation or decomposition of certain intermediates along the reaction path or for the rapid release of products during steady-state ATP hydrolysis (Gresser et al., 1982; Fellous et al., 1984). On the other hand, during either ATP synthesis by submitochondrial particles, the same formation or decomposition or release could be effected by conformation changes driven by the proton flux so that all three sites could function effectively. Although this artificially created high ATPase activity of mitochondrial ATP synthase probably has no importance in the intact natural system, it has surprisingly given us a deeper insight into the natural system by making it possible to show that internal rotation is not necessary for ATP synthesis driven by the proton flux.

ACKNOWLEDGMENT

The research described in this article was supported in part by a research grant from the National Institute of General Medical Sciences (GM 31463).

REFERENCES

Amzel, L.M., McKinney, M., Narayanan, P. and Pedersen, P.L. (1982) Proc. Natl. Acad. Sci. USA 79: 5852-5856.

Andrews, W.M., Hill, F.C. and Allison, W.S. (1984a) J. Biol. Chem. 259: 8219-8225.

Andrews, W.M., Hill, F.C. and Allison, W.S. (1984b) J. Biol. Chem. 259: 14378-14382.

Ferguson, S.J., Lloyd, W.J., Lyons, M.H. and Radda, G.K. (1975a) Eur. J. Biochem. 54: 117-126.

Fellous, G., Godinot, C., Baubichon, H., DiPietro, A. and Gautheron, D.C. (1984) Biochemistry 23: 5294-5299.

Ferguson, S.J., Lloyd, W.J. and Radda, G.K. (1975b) Eur. J. Biochem. 54: 127-133.

Gresser, M.J., Myers, J.A. and Boyer, P.D. (1982) J. Biol. Chem. 257: 12030-12038.

Runswick, M.J. and Walker, J.E. (1983) J. Biol. Chem. 258: 3081-3089.

Steinmeier, R.C. and Wang, J.H. (1979) Biochemistry 18: 11-18.

Sutton, R. and Ferguson, S.J. (1985a) Eur. J. Biochem. 148: 551-554.

Sutton, R. and Ferguson, S.J. (1985b) FEBS Lett. 179: 283-288.

Wang, J.H. (1984) Biochemistry 23: 6350-6354.

Wang, J.H. (1985) J. Biol. Chem. 260: 1374-1377.

Wang, J.H., Joshi, V. and Wu. J.C. (1986) Biochemistry 25: 7996-8001.

Wang, J.H., Cesana, J. and Wu, J.C. (1987) Biochemistry 26: in press.

Wu, J.C., Chuan, H. and Wang, J.H. (1987) J. Biol. Chem. 262: 5145-5150.

CATALYTIC AND NONCATALYTIC NUCLEOTIDE BINDING SITES OF

F$_1$-ATPases: PROBES OF LOCATION, STRUCTURE AND FUNCTION

BY USE OF 2-N$_3$-ATP

John G. Wise, Zhixiong Xue, Chad G. Miller, Brian J. Hicke, Karen J. Guerrero, Richard L. Cross*, and Paul D. Boyer

Molecular Biology Institute, University of California, Los Angeles, California 90034 and *Department of Biochemistry and Molecular Biology, SUNY, Health Sciences Center, Syracuse, New York 13210

INTRODUCTION

Some pertinent questions about the ATP synthase are: Where are the nucleotide binding sites on the enzyme? How many nucleotide binding sites participate directly in the catalytic reaction? Do catalytic sites change properties in a sequential manner during catalysis? Does occupancy of noncatalytic nucleotide binding sites affect or control catalytic events?

The approach taken in our current studies to these questions has been to probe catalytic and noncatalytic nucleotide binding sites with the photoreactive adenine-nucleotide analog, 2-N$_3$-ATP. We have reacted 2-N$_3$-ATP with F$_1$-ATPases isolated from spinach chloroplast, bovine heart mitochondrial and *Escherichia coli* membranes, and have identified the amino acid residues derivatized by the analog when bound specifically to catalytic or noncatalytic nucleotide binding sites. The results of these studies demonstrate similar nucleotide binding site structures on F$_1$-ATPases from all three sources and provides information on catalytic mechanism.

CATALYTIC AND NONCATALYTIC NUCLEOTIDE BINDING SITES: EXCHANGE WITH MEDIUM NUCLEOTIDES

Cross and Nalin (1982) defined two distinct classes of nucleotide binding sites based on a difference in the relative rates of exchange of F$_1$-bound nucleotide with nucleotides free in solution. They reported that of the six observed binding sites present on the enzyme, three exhibited little or no exchange with medium nucleotides during 1000 catalytic turnovers per MF_1. These sites were designated nonexchangeable sites, and were regarded as noncatalytic in character. The three sites that were readily exchangeable with medium nucleotide during turnover were regarded as potential catalytic sites. Wise et al. (1983) showed that *E. coli* F_1 (EF_1) also binds up to six mol of adenine nucleotides per mol of F_1 and that the same two classes of sites are present. It needs emphasis that exchange of F_1-bound nucleotide with medium nucleotide, not tightness of binding, differentiated noncatalytic and catalytic sites. Catalytic sites can bind nucleotide very tightly (Grubmeyer et al., 1982).

More recently, Xue et al. (1987a) have reported evidence that chloroplast F_1-ATPase (CF_1) like MF_1 and EF_1 has six nucleotide binding sites. They found that CF_1 could retain up to 4.5 mol of adenine nucleotide per mol of enzyme as measured by centrifugation through a Sephadex gel-filtration column. Three moles of nucleotide/mol CF_1 were resistant to exchange with medium nucleotide during catalytic turnover. These results suggest that CF_1 behaves similarly to MF_1 and EF_1 in regard to the presence of three noncatalytic binding sites. Further, as discussed below, they identified both a catalytic and a noncatalytic nucleotide binding site on individual beta subunits. This result, and the presence of three beta subunits per F_1-ATPase complex strongly supports the presence of six nucleotide binding sites on CF_1. Their observation of less than two exchangeable sites may indicate a faster off constant for this class of site on CF_1. It should be noted that MF_1 retained only about five mol of nucleotide per mol of enzyme when ATP was used in binding studies (Cross and Nalin, 1982). That the enzymes MF_1, EF_1 and CF_1 display similar stoichiometry and character of nucleotide binding is not surprising in view of the strong amino acid sequence homologies found for the nucleotide-binding subunits (Walker et al., 1985).

CATALYTIC AND NONCATALYTIC NUCLEOTIDE BINDING SITES: NUCLEOTIDE SPECIFICITY

It has long been recognized that catalytic sites on F_1-ATPases can readily bind and hydrolyze GTP, CTP, and ITP as well as ATP. Energy coupled reactions such as NTP-driven proton-translocation and oxidative or photophosphorylation exhibit a similar broad substrate specificity. In contrast, Harris et al. (1978) and Perlin et al. (1984) demonstrated that the noncatalytic binding sites on MF_1 and EF_1 were specific for adenine nucleotides. No tight binding to noncatalytic sites was detected with non-adenine nucleotides. Hence, a second experimental distinction between catalytic and noncatalytic nucleotide sites can be made on the basis of the nucleotide base specificity required for binding. Interestingly, the isolated alpha subunit of EF_1 contains a nucleotide binding site specific for adenine nucleotide (Dunn and Futai, 1980) that behaves similarly to the EF_1 noncatalytic site (see Dunn and Heppel, 1981; Senior and Wise, 1983).

β-SUBUNIT LOCATIONS OF CATALYTIC AND NONCATALYTIC NUCLEOTIDE BINDING SITES AS PROBED BY 2-N_3-ATP PHOTOLABELING

The photoactivatable substrate analog, 2-N_3-ATP has proven to be an exceptionally good reagent for probing the binding site locations in F_1-ATPases (Czarnecki et al., 1982). It acts as a good substrate (Abbott et al., 1984), and has been applied to CF_1, MF_1 and EF_1 in covalent incorporation studies. Abbott et al. (1984) showed that tightly bound 2-N_3-ADP and 2-N_3-ATP both labeled the same large beta subunit peptide of CF_1. Kironde and Cross (1986, 1987) showed that 2-N_3-ATP would bind to exchangeable or nonexchangeable noncatalytic sites in a manner analogous to the binding of naturally occurring adenine nucleotides. They further showed that different beta subunit peptides of MF_1 were labeled when the analog was covalently incorporated into the catalytic sites or noncatalytic sites.

Similar labeling of the beta subunit of F_1 from other sources has been shown. When CF_1 and EF_1 noncatalytic and catalytic site 2-N_3-ATP labeling patterns were examined by 2-D gel electrophoresis, CF_1 showed only beta subunit labeling (Xue et al., 1987a) while EF_1 samples showed mainly beta subunit labeling (90-95%) with slight detectable alpha subunit binding (Wise et al., 1987). When complete tryptic digests of 2-N_3-ATP-labeled MF_1, CF_1 and EF_1 samples were subjected to HPLC reversed phase chromatography, very distinct elution patterns for the noncatalytic versus catalytic site labeled peptides were evident (Cross et al., 1987; Xue et al., 1987a; Wise et al., 1987). Each of the F_1-noncatalytic site labeled peptides eluted earlier than the respective catalytic site peptides. These observations reinforce the idea that MF_1, CF_1 and EF_1 each contain separate catalytic and noncatalytic nucleotide binding sites composed in part by different regions of the beta subunit.

2-N_3-ATP LABELED BETA SUBUNIT PEPTIDES IN MF_1

Garin et al. (1986) identified one site and Cross et al. (1986, 1987) identified two sites in the MF_1 beta subunit primary sequence where covalent incorporation of 2-N_3-ATP occurs. When catalytic sites of MF_1 were covalently labeled with 2-N_3-[^{32}P]ATP, the majority of radioactivity was found associated with a single tryptic peptide. The sequencing of this peptide showed that beta-tyr-345 was derivatized. The noncatalytic sites on MF_1 were shown to be derivatized by $2-N_3$-ATP at tyrosine-368, in a tryptic-peptide adjacent to the catalytic site peptide. These results suggest a relatively close proximity of the two different types of sites on the enzyme. It is of interest that the nucleotide analog FSBI reacts with the catalytic site tyrosine (Bullough and Allison, 1987), and that FSBA reacts with the noncatalytic site tyrosine. FSBA was originally thought to react at catalytic sites (see Bullough and Allison, 1986), but this reaction occurs at what are now known to be noncatalytic sites (Cross et al., 1987).

CATALYTIC SITES IN CF_1 AND EF_1 ARE SPECIFICALLY LABELED BY $2-N_3$-ATP

In recent work, Xue et al. (1987b) have shown that beta-tyr-362 is derivatized when catalytic sites on CF_1 are photolabeled by $2-N_3$-ATP. In $2-N_3$-ATP labeling studies of the EF_1 catalytic site, Wise et al. (1987) showed that the majority of incorporated analog was found associated with two purified HPLC peptides (40 and 45% of total cpms incorporated). The first catalytic site peptide, when subjected to sequence analysis, indicated that $2-N_3$-ATP labeled beta-tyr-331 of EF_1. The second peptide was resistant to Edman degradation as no detectable PTH-amino acid was observed in any of the automated cycles (analysis of up to 500 pmols of radiolabeled peptide on a gas-phase microsequencer). A parallel comparison of the amino acid compositions (acid hydrolysis) of the two peptides suggests that they are identical. The difference in HPLC elution time and susceptibility to Edman degradation between the two peptides may reflect the cyclization of the N-terminal glutamine residue of one of the peptides. Labeling data for the highly conserved catalytic site peptides of the three enzymes are summarized in Fig. 1.

NONCATALYTIC SITES IN CF_1 AND EF_1 ARE ALSO SPECIFICALLY LABELED BY $2-N_3$-ATP

The $2-N_3$-ATP-labeled noncatalytic site tryptic peptide of CF_1 (Xue et al., 1987a) and the homologous noncatalytic site peptide labeled in EF_1 (Wise et al., 1987) have been sequenced. Replicate samples of peptides from both sources (purified by ion exchange and reversed phase HPLC) showed that tyrosine-385 of the CF_1 beta subunit and tyrosine-354 of the EF_1 beta subunit were derivatized by the analog. Fig. 1 summarizes the present data and shows that the catalytic and noncatalytic site peptides of MF_1, CF_1 and EF_1 labeled by $2-N_3$-ATP and $2-N_3$-ADP are highly conserved homologous peptides with principal derivatization on homologous tyrosyl residues.

The ability to distinguish between catalytic and noncatalytic site binding, and between incorporated ADP- and ATP-moieties, should serve to help research efforts aimed at understanding the ATP synthases.

ALPHA SUBUNIT DERIVATIZATION BY NUCLEOTIDE ANALOGS

The sites of covalent incorporation of nucleotide analogs into F_1-ATPase alpha subunits are unfortunately not commonly known, and to date only amino acid residues modified in the beta subunit have been determined. That the alpha subunit has been implicated in forming nucleotide binding structures in F_1 stems from observations of incorporation of radioactive analog into alpha subunit bands on SDS gels, and in studies performed on isolated alpha subunits from bacterial sources (see Ohta et al., 1980; Dunn and Futai, 1980). As mentioned above, it has been suggested that the alpha subunit of F_1 may contribute to the nucleotide binding site structure(s) in F_1. It may be of interest to analyze the amino acid sequences of minor components of the EF_1 samples when labeled with 2-N_3-ATP, since such derivatization results in reproducible alpha subunit labeling.

HOW MANY CATALYTIC SITES ARE ACTIVE DURING NET CATALYSIS?

Although the binding change mechanism with cooperative participation of at least two catalytic sites has gained considerable support (Gresser et al., 1982; Cross et al., 1982; Lubben et al., 1984; Futai and Kanazawa, 1983; Senior and Wise, 1983; and Hatefi, 1985), uncertainty remains as to whether only a single catalytic site exists that is modulated by one or more regulatory sites (Bruist and Hammes, 1981; Wang, 1984; Williams et al., 1987). Direct evidence for more than one active catalytic site per F_1 came originally from the work of Rosen et al. (1979) in their studies of hexokinase inaccessible ATP bound to CF_1 during steady-state photophosphorylation. Grubmeyer and Penefsky (1981a,b) have provided additional evidence for at least two active catalytic sites per MF_1. These researchers measured the binding and hydrolysis of TNP-ATP at one and two nucleotide binding sites on MF_1. More recently Grubmeyer et al. (1982) and Cross et al. (1982) provided kinetic evidence for the possible participation of three catalytic sites. This more recent work has been extended to the EF_1-ATPase (Wise et al., 1984), suggesting the general application of the structures and mechanism.

Catalytic 2-Azido-ATP Labeled Beta Subunit Peptides:

F₁		Residue Modified *	Peptide Labeled [a]
Bovine Mitochondria	[b,c]	Tyr-345	A I A E L G I Y* P A V D P L D S T S R
Spinach Chloroplast	[d]	Tyr-362	G I Y* P A V D P L D S T S T M L Q P R
Escherichia coli	[e]	Tyr-331	Q I A S L G I Y* P A V D P L D S T S R

Noncatalytic 2-Azido-ATP Labeled Beta Subunit Peptides:

F₁		Residue Modified *	Peptide Labeled [a]
Bovine Mitochondria	[c]	Tyr-368	I M N P N I V G S E* H Y D V A R
Spinach Chloroplast	[d]	Tyr-385	I V G E E H Y* E I A Q R
Escherichia coli	[e]	Tyr-354	Q L D P L V V G Q E* H Y D T A R

* Residue number refers to the site of 2-Azido-ATP derivatization counted from the N-terminus of the respective beta subunit.

[a] Boxes indicate conserved residues from Walker et al, 1985.
[b] From Garin et al, 1986.
[c] From Cross et al, 1986,1987.
[d] From Xue et al, 1987b.
[e] From Wise et al, 1987.

Fig. 1.

STATIC ASYMMETRY VERSUS APPARENT SYMMETRY DURING CATALYSIS AND IMPLICATIONS FOR ROTATIONAL MECHANISMS

An important extension of the binding change mechanism is that the three beta subunits of F_1 react in a coordinated sequence such that at any one instant during catalysis, or when the enzyme is catalytically inactive, all of the beta subunits exist in distinctly different conformations. During catalysis, all of the beta subunits would pass sequentially through identical conformations. The three beta subunit conformations, when observed over times greater than the reciprocal of the enzymes turnover number, would appear to be identical. This hypothesis has been tested (Melese et al., 1987) by measurement of the reactivity of beta subunits with general chemical derivatizing reagents (succinic anhydride and iodoacetate). Results of these studies indicate that turnover of the enzyme induces the averaging of chemical reactivity of all beta subunits, while treatment without turnover suggests unique beta subunit conformations, as predicted by the binding change mechanism.

The concept of static asymmetry is useful in explaining results suggesting divalent metal-ion binding asymmetry. It should be noted that F_1 during catalysis has a Mg^{++} stoichiometry of at least 2 per F_1 (Senior et al., 1980), whereas when incubated with a strong metal ion chelator, F_1 can be stripped of all but one divalent metal ion. Further removal of metal ion results in irreversible protein precipitation (Senior, 1979; Williams et al., 1987). The apparently asymmetric "structural Mg^{++}" observed under strong chelation/no turnover conditions should exist during catalysis, thereby arguing against equivalent beta subunit conformations during catalysis. However, it is quite consistent with the binding change mechanism, discussed above, for the enzyme to display static asymmetry of beta subunits when no catalytic turnover is occurring, or of permanent asymmetry of the "noncatalytic core" subunits (see Gresser et al., 1982). The "asymmetric metal" binding site has been found associated with alpha and gamma subunits of F_1 (see Williams et al., 1987). These subunits were hypothesized by Gresser et al. (1982) to be the major components of the noncatalytic core of the enzyme. That the "noncatalytic core" of MF_1 may display permanent asymmetry has been recently suggested by Kironde and Cross (1987).

The $\alpha_3\beta_3\gamma\delta\epsilon$ structural subunit arrangement of F_1-ATPases, and the mechanistic need to explain static asymmetry while apparent symmetry is observed during catalysis led to the postulation that beta subunits may rotate relative to the other F_1 subunits during catalytic turnover (Gresser et al., 1982). More recently, others have postulated rotational mechanisms (Cox et al., 1984; Mitchell, 1985). The mechanisms suggested by Boyer and Cox explain uniform interaction of the beta subunits with noncatalytic subunits. At the present time, while we favor a rotational catalytic mechanism as the simplest means of accommodating asymmetry in the absence of turnover despite apparent beta subunit symmetry during catalysis, it is obvious that considerably more research is needed before rotational mechanisms can either be fully accepted or fully denied.

NONCATALYTIC NUCLEOTIDE BINDING SITES: DO THEY CONTROL CATALYSIS?

This has long been an active area of interest, but thus far no clear cut answer has been obtained. A number of laboratories have attributed a strong inhibition of hydrolytic activity to ADP binding to noncatalytic sites. For instance, Fellous et al. (1984) have obtained evidence for the modulation of ATP hydrolytic activity of pig heart MF_1 when ADP is bound to noncatalytic site(s) on the enzyme. Wise and Senior (1985) were unable to detect any regulatory effect of occupancy of noncatalytic sites on either oxidative phosphorylation or hydrolytic activities of the *E. coli* enzyme. The observations of Hammes suggest a noncatalytic control site on CF_1 (Bruist and Hammes, 1981; Leckband and Hammes, 1987). One of us (Z.X., unpublished) has shown that the site described by Hammes, when labeled by $2-N_3$-ATP, derivatizes beta subunit tyrosine-385 of CF_1 (noncatalytic site).

In summary, modulation of ATPase activity in MF_1 and CF_1 has been observed but the relevance to physiological control is uncertain. No experimental evidence for control of catalysis by nucleotide bound at noncatalytic sites on EF_1 has yet been reported. The possibility of control of catalysis by nucleotide bound at noncatalytic sites obviously needs further evaluation.

ACKNOWLEDGMENTS

Supported by USPHS grants GM11094 to PDB and GM23152 to RLC. JGW was supported by an American Heart Association Senior Investigatorship (Greater Los Angeles Affiliate). CGM was supported in part by USPHS National Research Service Award GM07165.

REFERENCES

Abbott, M.S., Czarnecki, J.J., and Selman, B.R. (1984) J. Biol. Chem. 259, 12271-12278.
Bruist, M.F. and Hammes, G.G. (1981) Biochemistry 20, 6298-6305.

Bullough, D.A. and Allison, W.S. (1986) J. Biol. Chem. 261, 5722-5730.

Bullough, D.A. and Allison, W.S. (1987) J. Biol. Chem. 261, 14171-14177.

Cox, G.B., Dans, D.A., Fimmel, A.L., Gibson, F. and Hatch, L. (1984) Biochim. Biophys. Acta 768, 201-208.

Cross, R.L. and Nalin, C.M. (1982) J. Biol. Chem. 257, 2874-2881.

Cross, R.L., Cunningham, D., Miller, C.G., Xue, Z., Zhou, J.M. and Boyer, P.D. (1987) Proc. Natl. Acad. Sci. USA, in press.

Cross, R.L., Kironde, F.A.S. and Cunningham, D. (1986) EBEC Reports 4, 411.

Cross, R.L., Grubmeyer, C. and Penefsky, H.S (1982) J. Biol. Chem. 257, 12101-12105.

Czarnecki, J.J., Abbott, M.S. and Selman, B.R. (1982) Proc. Natl. Acad. Sci. USA 79, 7744-7748.

Dunn, S.D. and Futai, M. (1980) J. Biol. Chem. 255, 113-118.

Dunn, S.D. and Heppel, L.A. (1981) Arch. Biochem. Biophys. 210, 421-436.

Fellous, G., Godinot, C., Baubichon, H., DiPietro, A. and Gautheron, D.C. (1984) Biochemistry 23, 5294-5299.

Futai, M. and Kanazawa, H. (1983) Microbiol. Rev. 47, 285-312.

Garin, J., Boulay, F., Issartel, J.P., Lunardi, J. and Vignais, P.V. (1986) Biochemistry 25, 4431-4437.

Gresser, M.J., Myers, J.A., and Boyer, P.D. (1982) J. Biol. Chem. 257, 12030-12038.

Grubmeyer, C. and Penefsky, H.S. (1981a) J. Biol. Chem. 256, 3718-3727.

Grubmeyer, C. and Penefsky, H.S. (1981b) J. Biol. Chem. 256, 3728-3734.

Grubmeyer, C., Cross, R.L. and Penefsky, H.S. (1982) J. Biol. Chem. 257, 12092-12100.

Harris, D.A., Gomez-Fernandez, J.C., Klungsoyr, L. and Radda, G.K. (1978) Biochim. Biophys. Acta 504, 364-383.

Hatefi, Y. (1985) Annu. Rev. Biochem. 54, 1015-1069.

Kironde, F.A.S. and Cross, R.L. (1986) J. Biol. Chem. 261, 12544-12549.

Kironde, F.A.S. and Cross, R.L. (1987) J. Biol. Chem. 262, 3488-3495.

Leckband, B. and Hammes, G.G. (1987) Biochemistry, in press.

Lubben, M, Lucken, U. , Weber, J., and Schafer, G. (1984) Eur. J. Biochem. 143, 483-490.

Melese, T., Crans, D.C. and Boyer, P.D. (1987) in preparation.

Mitchell, P. (1985) FEBS Lett. 182, 1-7.

Ohta, S., Tsuboi, M., Oshima, T., Yoshida, M. and Kagawa, Y. (1980) J. Biochem. 87, 1609-1617.

Perlin, D.S., Latchney, L.R., Wise, J.G. and Senior, A.E. (1984) Biochemistry 23, 4998-5003.

Rosen, G, Gresser, M., Vinkler, C. and Boyer, P.D. (1979) J. Biol. Chem. 254, 10654-10661.

Senior, A.E (1979) J. Biol. Chem. 254, 11319-11322.

Senior, A.E. and Wise, J.G. (1983) J. Membr. Biol. 73, 105-124.

Senior, A.E., Richardson, L.V., Baker, K. and Wise, J.G. (1980) J. Biol. Chem. 255, 7211-7217.

Walker, J.E., Fearnley, I.M., Gay, N.J., Gibson, B.W., Northrop, F.D., Powell, S.J., Runswick, M.J., Saraste, M. and Tybulewicz, V.L.J. (1985) J. Mol. Biol. 184, 677-701.

Wang, J.H. (1984) Biochemistry 23, 6350-6355.

Williams, N., Hullihen, J.M. and Pedersen, P.L. (1987) Biochemistry 26, 162-169.

Wise, J.G. and Senior, A.E. (1985) Biochemistry 24, 6949-6954.

Wise, J.G., Duncan, T.M, Latchney, L.R., Cox, D.N. and Senior, A.E. (1983) Biochem. J. 215, 343-350.

Wise, J.G., Latchney, L.R., Ferguson, A.M. and Senior, A.E. (1984) Biochemistry 23, 1426-1432.

Wise, J.G. et al. (1987) in preparation.

Xue, Z., Zhou, J.-M., Melese, T., Cross, R.L. and Boyer, P.D. (1987a) Biochemistry, in press.

Xue, Z. et al. (1987b) in preparation.

INTERACTION OF ARYLAZIDO-β-ALANYL ATP[a] WITH THE ATPase ENZYME

OF *RHODOSPIRILLUM RUBRUM* CHROMATOPHORES

Richard John Guillory

John A. Burns School of Medicine, Department of Biochemistry
and Biophysics, University of Hawaii, Honolulu, Hawaii 96822

ABSTRACT

The ATP photoaffinity probe arylazido-β-alanyl ATP is able to inhibit the ATP dependent reactions of the *Rhodospirillum rubrum* chromatophore membrane in a specific and light dependent manner. Following the light induced photoprobe interaction with LiCl treated chromatophores (resolved of the ATPase β subunit), the LiCl dialyzed supernatant (i.e., β subunit) is still capable of reconstituting photophosphorylation as in the non-photoprobe treated membrane. However, the ability of the β subunit preparation to reconstitute membrane bound ATPase activity is inhibited.

INTRODUCTION

The first irreversible photoaffinity labeling of the nucleotide binding sites of the beef heart mitochondrial ATPase was reported in 1976 (Russell et al., 1976) utilizing the photoprobe arylazido-β-alanyl ATP (Jeng and Guillory, 1975). This was followed almost simultaneously by the report of Lunardi et al. (1977) of the use of a similar probe N-4-azido-2-nitrophenyl amino butyryl-ADP. The work by Lunardi et al. (1977) and the subsequent work of Cosson and Guillory (1979) showed that both nucleotide analogues bind irreversibly to both α and β subunits of the proton translocating ATPase enzyme. In the case of the mitochondrial F_1 photoprobe covalent interaction

---

[a]The abbreviations used are: arylazido-β-alanyl ATP; A3'-O-3-[N-(4-azido-2-nitrophenyl)amino]propionyl ATP.

Molecular Structure, Function, and Assembly of the ATP Synthases
Edited by Sangkot Marzuki
Plenum Press, New York

of arylazido-β-alanyl ATP is associated with the binding of a mole of analogue per mole of enzyme with a fairly even distribution of the mole between both α and β subunits.   This observation led Cosson and Guillory (1979) to postulate as one possibility that the active ATPase site for the F_1 enzyme might be composed of a portion of an α and an adjacent β subunit of the enzyme (see as well Lubben et al., 1984).

The energy conserving ATPase enzyme of R. *rubrum* represents an enzyme complex resembling in a number of ways that of mitochondrial and chloroplast membranes.   The first successful resolution of the chromatophore membrane of its ATPase activity and ATP dependent reactions was accomplished by Fisher and Guillory (1969).   At that time it was shown that extraction of the chromatophore membrane with buffered 2 M LiCl resulted in complete loss of both ATPase activity and photophosphorylation.   A number of other energy dependent reactions independent of an ATP requirement such as pyrophosphatase, light driven transhydrogenation and photodependent pyrophosphate synthesis, were reported not to be affected. Subsequently it was demonstrated by Philosop et al. (1977) that in the presence of ATP there is released by such LiCl extraction protein factor capable of reconstituting both photophosphorylation and Mg dependent ATPase activity to the resolved membrane. The protein factor was identified as being the β subunit of the ATPase enzyme complex. The presence of ATP during extraction appears to stabilize the β subunit as it is selectively released from the membrane.

The purpose of the present study was to attempt to use the photoprobe arylazido-β-alanyl ATP to answer the question as to whether the photophosphorylation of ADP (an energy conserving state) and the hydrolysis of ATP (an energy requiring state) are independent functions of the α or β subunit or are functions of a required combined action.   The approach utilized is the photoirradiation of LiCl treated chromatophores in the presence of arylazido-β-alanyl ATP resulting in the specific labeling of the α ATPase subunit. Evaluation of the potential for reconstitution of photophosphorylation and ATPase activities in the Photoprobe treated preparation, by a dialyzed LiCl extract (β-subunit) is then carried out and compared to a control no photoirradiated preparation.

MATERIALS AND METHODS

Chromatophores were prepared as previously described (Fisher and Guillory, 1971) as was the assay of ATPase activity (Fisher and Guillory, 1967).

Photophosphorylation of ADP to ATP was measured as the formation of glucose-6-phosphate in the presence of glucose and hexokinase. The standard medium contained 16 μg BChl (chromatophores) in 3 ml of 30 mM Tricine pH 8, 3 mM $MgCl_2$, 18 mM glucose, 150 μg hexokinase, 5 mM ADP, 5 mM K_2HPO_4 and 68 μM phenazine methosulfate. This was incubated for 10 min in Warburg vessels (5 ml) subjected to constant agitation at 30°C. Irradiation was carried out through the bath by a bank of 30 watt General Electric reflector lamps 8 cm from the bottom of the reaction vessels. The reaction was terminated by tipping in from the side arm 0.3 ml of (70%) perchloric acid. Two ml of the protein free supernatant following centrifugation was neutralized (KOH) to pH 7.5, the neutralized solution frozen, thawed and centrifuged. An aliquot of the neutralized supernatant was assayed enzymatically for glucose-6-phosphate in a total volume of 3 ml containing 50 mM Triethanolamine pH 7.5, 1.1 mM $NADP^+$, 10 mM $MgCl_2$ and 1.5 μg glucose-6-phosphate dehydrogenase (Sigma Chemical Co).

LiCl particles were prepared as described by Fisher and Guillory (1969) with the modification that the extraction medium contain as well 4 mM ATP (Philosop et al., 1977). The LiCl supernatant was dialyzed for at least 18 hrs at 4°C against two changes of 0.05 M Tricine NaOH, pH 7.6, 1 mM ATP, 4 mM $MgCl_2$ and is termed dialyzed coupling factor.

Reconstitution of photophosphorylation. In the reconstitution step depleted chromatophores (15 μg BChl) were incubated in Warburg flasks at 30° in 0.4 ml 30 mM Tricine NaOH pH 7.6, 25 mM $MgCl_2$, 4 mM ATP, 18 mM glucose with varying quantities of dialyzed coupling factor. Following a 30 min dark incubation a photophosphorylation mixture is added such that a final volume of 3 ml contains 30 mM Tricine, 5 mM ADP, 5 mM K_2HPO_4 and 66 μM phenazine methosulfate. The mixture was incubated at 30° with constant agitation and light irradiation. At 10 min the reaction was terminated by tipping in 0.3 ml of 70% perchloric acid. Neutralization and analysis of glucose-6-phosphate was carried out as described above.

Reconstitution of ATPase. Resolved chromatophores (15 μg BChl) were incubated with varying quantities of dialyzed coupling factor in a total volume of 0.4 ml containing 30 mM Tricine NaOH pH 7.6, 25 mM $MgCl_2$, 4 mM ATP, 18 mM sucrose. After 30 min at 37°, 0.5 ml of a solution is added containing 66 mM glycylglycine pH 8, 196 mM sucrose, 4 mM $MgCl_2$, 6 mM phosphoenol pyruvate and 0.005 mM carbonyl cyanide phenylhydrazine together with 50 μg of pyruvate kinase and ATP to a final concentration of 2 mM. The total volume of 1 ml was incubated at 30° for 10 min and the reaction terminated by the addition of 0.2 ml of (10 N) perchloric acid. An aliquot of the protein-free supernatant (0.6 ml) was assayed for inorganic phosphate.

Photolabeling with arylazido-β-alanyl ATP was carried out by incubation of the photoprobe with 15 μg BChl in 0.2 ml Tris-Cl pH 7.0, photoirradiation being accomplished with 40 flashes (or less) 10 cm from the light source as described by Cosson and Guillory (1979).

Coupling factor ATPase protein from *R. rubrum* was extracted from an acetone powder of *R. rubrum* chromatophores and purified by $(NH_4)_2SO_4$ fractionation, sepharose 6B chromatography, and ultrafiltration (Johansson, 1973). Acrylamide gel electrophoresis was carried out as described by Davis (1964) or Weber and Osborn (1969).

RESULTS

The soluble *R. rubrum* enzyme when electrophoresed on 7% polyacrylamide gels consists of a single major protein with a number of minor contaminating peptides. When subjected to electrophoresis in sodium dodecylsulfate on 10% acrylamide gels the banding pattern observed was similar to that reported for other energy transducing ATPase enzymes and consistent with the reported subunit molecular weight of α (54 kDa), β (50 kDa), γ (32 kDa), δ (13 kDa) and ε (7.5 kDa). The molecular weight of the *R. rubrum* enzyme has been reported to be 350,000 (Johansson and Baltscheffsky, 1975).

Table I demonstrates the extreme sensitivity of the soluble enzyme to photoirradiation. In the absence of a filter to limit UV radiation, 10 flashes are sufficient to decrease the enzymatic activity 96% in the absence of the photoprobe. In the presence of a glass shield there is still a 21% decrease in enzymatic activity. Although it is not possible to show a photodependent inhibitory influence of the *R. rubrum* ATPase a photo-dependency for covalent labeling is readily demonstrated. Fig. 1 shows that the labeling pattern of the *R. rubrum* enzyme is more specific than that observed with the heart mitochondrial ATPase. With the mitochondrial F_1, photoprobe interaction is associated with the binding of a mole of analogue per mole of enzyme with a fairly even distribution of the one mole between both α and β subunits. In the case of the soluble *R. rubrum* enzyme a similar inhibition binding stoichiometry is observed except that covalent association of the nucleotide probe is primarily with the α subunit of the enzyme. In contrast to the soluble enzyme the chromatophore bound ATPase activity as well as photophosphorylation can readily be shown to be inhibited by arylazido-β-alanyl ATP in a photodependent manner (Table II).

Table I. Photodependent Inhibition of *R. rubrum* F_1-ATPase Activity by Arylazido-β-Alanyl ATP.

	Control	Arylazido-β-Alanyl ATP
Nonirradiated	190	18
Photoirradiated	7	12
Photoirradiated (with glass shield)	149	11

R. rubrum ATPase (16 μg), purified to the Sepharose 6B chromatography step (Johansson et al. 1973) was photoirradiated [at 0-4°, 10 flashes, 10 cm from the light source (Cosson and Guillory (1978)] in 200 μl of 25 mM Tris-Cl pH 7 with or without 216 μM arylazido-β-alanyl ATP. Ca^{++} dependent ATPase is given as μmoles of phosphate released (mg protein)$^{-1}$hr^{-1}.

Chromatophores were resolved of ATPase and photophosphorylation activity by treatment with LiCl-ATP as described in the Methods section and the LiCl-ATP extract used to reconstitute both activities in the chromatophore membrane.

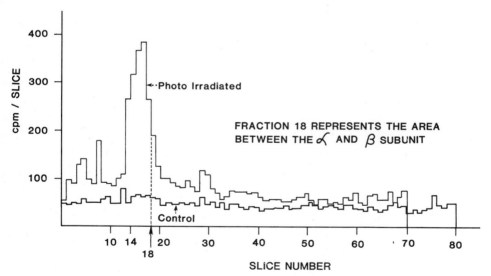

Fig. 1. Labeling of Soluble *R. rubrum* F_1-ATPase by Arylazido-β-Alanyl ATP. Following photoirradiation in the presence of 200 mM arylazido-β-[^{3}H]alanyl ATP [10 flashes, 10 cm from the light source as described by Cosson and Guillory (1979)], the solution was applied to a 10% polyacrylamide sodium dodecylsulfate gels for electrophoresis according to the procedure described by Weber and Osborn (1969). The gels, stained with coomassie blue, were scanned at 550 nm and then sliced with a multiblade slicer. Individual slices were digested overnight with 0.3 ml of 30% H_2O_2 at 65° and the radioactivity of each sample measured by liquid scintillation in 5 ml aquasol (New England Nuclear).

233

Table II. Influence of Arylazido-β-Alanyl ATP on Photophosphorylation and Mg^{++}-ATPase Activity of Chromatophores of *R. rubrum*

	ATPase	% Control	Photo- phosphorylation	% Control
Control	553	100	651	100
Arylazido-β-alanyl ATP	523	95	705	108
Arylazido-β-alanyl ATP Photoirradiated	344	62	205	31

Chromatophores (9.21 μg BChl) were incubated at 0-4° in 200 μl 25 mM Tris-Cl pH 7.0 with or without 203 μM arylazido-β-alanyl ATP. The preparation was photoirradiated with 40 flashes, 20 cm from the light source (Cosson and Guillory, 1978). ATPase activity is recorded as μmoles phosphate released (mg BChl)$^{-1}$hr^{-1}. Photophosphorylation activity is recorded as μmoles ATP synthesized (mg BChl)$^{-1}$hr^{-1}.

Maximum reconstitution of photophosphorylation (80%) and ATPase activity (65%) occurred at a level of 75 μg protein (dialyzed LiCl extract) with 15 μg BChl LiCl particles.

The LiCl-ATP resolved chromatophores were subjected to photoirradiation in the presence of varying arylazido-β-alanyl ATP concentrations using conditions effective in the irreversible inhibition of photophosphorylation and membrane bound ATPase activity. The resolved and photoirradiated arylazido-β-alanyl ATP treated chromatophores were then subjected to reconstitution studies using LiCl extracts (i.e., β subunit). Fig. 2 shows that the photoprobe pretreated LiCl particles still retained the ability of having photophosphorylation reconstituted with the dialyzed coupling factor while the membrane bound ATPase reconstitution was inhibited proportionately to the concentration of arylazido-β-alanyl ATP used during the prereconstitution photoirradiation phase.

CONCLUSION

The labeling pattern of *R. rubrum* ATPase by arylazido-β-alanyl ATP appears to be more specific than that obtained with the mitochondrial F_1-ATPase. This difference may represent a difference in subunit organization for the two enzymes. With mitochondrial F_1, inhibition of ATPase activity by the photoprobe is associated with the binding of a mole of analogue per mole of enzyme with binding fairly evenly distributed between both α and β subunits. In the case of the soluble *R. rubrum* enzyme a similar inhibition binding-stoichiometry is observed except that covalent association of the nucleotide probe is almost exclusively with the α subunit of the enzyme. Perhaps the cooperative nucleotide activity observed for the mitochondrial ATPase is not present in the bacterial F_1 which unlike the mitochondrial ATPase is an enzyme associated with the cytoplasmic side of a plasma membrane.

In considering subunit functions for the energy transducing ATPase enzymes one must be cautious in ascribing a common role to the α and β subunit of these different enzymes. The distinction between α and β is quite arbitrarily based only upon their relative molecular weights. There is in fact no presently established fundamental physiological role for the distinction. It is conceivable that a functional requirement for the α subunit of the *R. rubrum* enzyme may be identical to that for the β subunit for the mammalian mitochondrial enzyme.

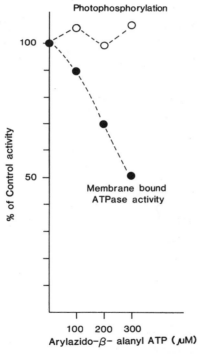

Fig. 2. Reconstitution of Photophosphorylation and Mg^{++} Dependent ATPase Activity in Resolved Chromatophores Previously Photoirradiated in the Presence of Arylazido-β-Alanyl ATP. Photoirradiation of resolved chromatophores (20 μg BChl) at 25° took place in 200 μl of 25 mM Tris pH 7 with varying arylazido-β-alanyl ATP concentrations using 10 flashes at a 10 cm distance from the light source. Reconstitution with 15 μg BChl was effected using a dialyzed LiCl extract containing 93 μg of protein. The control reconstituted ATPase activity was 535 μmol phosphate released $(mg\ BChl)^{-1}hr^{-1}$ and the reconstituted photophosphorylation activity, 460 μmol ATP synthesized $(mg\ BChl)^{-1}hr^{-1}$.

When *R. rubrum* chromatophores are treated with LiCl in the presence of ATP the "β" subunit is resolved from the membrane; consequently interaction of the resolved membrane with arylazido-β-alanyl ATP would be expected, following photoirradiation, to result in covalent labeling of the α subunit of the enzyme which is still present on the membrane. The experimental observation that such an interaction results in the photodependent inhibition of the reconstitution of membrane bound ATPase activity but has no influence on the ability of the coupling factor preparation to reconstitute photophosphorylation could be taken to indicate

(a)   that the reconstitution factor (i.e., "β" subunit) is specifically required for the phosphorylation reaction,

(b)   that an intact α subunit is required for the catalysis of membrane-bound ATPase activity and

(c)   that both α and β have independent physiological functions with respect to the energy transducing requirements of the membrane system.

Perhaps the nucleotide binding sites on both α and β subunits are catalytically active with the specificity of activity being dependent upon the particular physiological condition with respect to energy conservation or energy utilization.   Consequently in *R. rubrum* the ATPase reaction and the ATP synthase pathway usually considered to be reversible pathways of a single subunit might be best represented as:

$$\text{ATP} + \text{HOH} \xrightarrow{\quad \alpha \quad} \rightleftharpoons \text{ADP} + \text{P}_i$$
$$\text{ATP} + \text{HOH} \underset{\beta}{\xleftarrow{\qquad\qquad}} \rightleftharpoons \text{ADP} + \text{P}_i$$

Current work is directed towards a further evaluation of the function of the α and β subunits of the *R. rubrum* ATPase for the energy conserving and energy dissipative profile of the chromatophore membrane.

REFERENCES

Cosson, J. and Guillory, R.J. (1979) J. Biol. Chem. 254, 2946-2955.

Davis, B.J. (1964) Annals N.Y. Acad. Sci. 121, 404-427.

Fisher, R.R. and Guillory, R.J. (1969) FEBS Lett. 3, 27-30.

Fisher, R.R. and Guillory, R.J. (1971) J. Biol. Chem. 246, 4677-4686.

Fisher, R.R. and Guillory, R.J. (1967) Biochim. Biophys. Acta 143, 654-656.

Jeng, S.J. and Guillory, R.J., (1975) J. Supramol. Struct. 3, 448-468.

Johansson, B.C. (1973) Eur. J. Biochem. 40, 109-117.

Johansson, B.C. and Baltscheffsky, M. (1975) FEBS Lett. 53, 221-224.

Lubben, M., Luken, U., Weber, J. and Schafer, G. (1984) Eur. J. Biochem. 143, 483-490.

Lunardi, J., Lauquin, G.J.M. and Vignais, P.V. (1977) FEBS Lett. 80, 317-323.

Philosop, S., Bender, A. and Gromet-Elhanan, Z. (1977) J. Biol. Chem. 252, 8747-8752.

Russell, J., Jeng, S.J. and Guillory, R.J. (1976) Biochem. Biophys. Res. Com.. 70, 1225-1234.

Weber, K. and Osborn, M. (1969) J. Biol. Chem. 244, 4406-4412.

CATALYTIC COOPERATIVITY IN F_1-ATPase:

PHOTOAFFINITY LABELING STUDIES WITH BzATP[a]

Peter S. Coleman and Sharon Ackerman[b]

Laboratory of Biochemistry, Dept. of Biology, New York University
Washington Square, New York, N.Y. 10003

INTRODUCTION

In 1982, Williams and Coleman (1982) introduced the photoaffinity label BzATP, (3'-o-[4-Benzoyl]Benzoyl Adenosine 5'-Triphosphate), in studies with the rat liver mitochondrial F_1-ATPase. A principal argument favoring the use of benzophenone-derivative photoaffinity labels is the very low probability of reaction between their excited state and water, the biochemical solvent. This situation contrasts dramatically with nitrene-generating photoprobes. The conclusions drawn from these initial studies emphasized that benzophenone-adenine nucleotide adducts could serve as informative photoactivatable substrate analogs for the nucleotide binding sites of this complex enzyme. With rat liver F_1, BzATP appeared to behave both as a fair substrate and also as a competitive inhibitor for ATPase activity. Under actinic illumination, the probe bound covalently to the F_1 in a site-specific manner, which was preventable by the presence of ATP. Furthermore, BzATP and BzADP were found, depending on experimental conditions, to seek out nucleotide-specific sites on the α and/or the β subunits of the enzyme. Upon light-activated, site-specific covalent binding to the rat liver F_1, ATPase activity was lost.

Since the first report on the use of BzATP as a photoaffinity probe for nucleotide binding domains, other laboratories have used it successfully with a variety of ATP-requiring enzymes (Mahmood and Yount, 1984; Manolson et al., 1985), including the chloroplast F_1 (Bar-Zvi and Shavit, 1984; Kambouris and Hammes,

---

[a]Supported by a grant from the USPHS (GM 36619) to PSC
[b]Present address: Dept. of Biological Sciences, Columbia University, New York, N.Y. 10027

1985). However, the majority of mechanistic studies on mammalian F_1 have come from laboratories that employ the beef heart mitochondrial enzyme, and it is from such work that the principal evidence for catalytic cooperativity has been most forthcoming (Grubmeyer et al., 1982; Cross et al., 1982; Penefsky, 1985a; Penefsky, 1985b; Gresser et al., 1982). Thus, we have now examined the effects of Bz-adenine nucleotides on the catalytic properties of the beef heart enzyme (BF_1), paying particular attention to evidence that might comment on a mechanism involving catalytic cooperativity for ATP hydrolysis. These studies required the synthesis of [^{3}H]BzATP (^{3}H in the benzophenone ring), and [γ-^{32}P]BzATP. Beef heart F_1 was isolated and purified according to standard procedures (Penefsky, 1979), slightly modified.

KINETICS

With the rat liver F_1, BzATP had been found to be a rapidly hydrolyzable substrate, displaying a specific activity about 10% that of ATP, and an apparent K_m nearly identical with that of the natural nucleotide (Williams and Coleman, 1982). The rapid turnover of the rat liver enzyme with BzATP readily permitted spectrophotometric kinetics analysis by means of the standard coupled enzyme assay. With BF_1, however, catalytic turnover with BzATP was considerably lower, necessitating a ^{32}Pi release assay with [^{32}P]BzATP, and via this approach the kinetic constants given in Table I were obtained. It is apparent that although the K_m for

Table I. Kinetic Constants for F_1 Hydrolysis of ATP and BzATP

Substrate	K_m (μM)	Vmax (U/mg)	kcat (s^{-1})	kcat/K_m (10^6 x $M^{-1}s^{-1}$)
ATP	140	45	270	2.0
BzATP	0.53	0.19	1.14	2.2

The reaction mixture in a volume of 1 ml contained: 50 mM Tris–acetate, pH 8.0, 1 mM MgCl$_2$, 200 ng F_1, and a range of concentrations of either [γ-2P]ATP (0.05-0.33 μM) or [γ-^{32}P]BzATP (0.23-1.44 μM). ATPase activity was determined by analysis of ^{32}Pi released. For each concentration of either [γ-^{32}P]ATP or [γ-^{32}P]BzATP, the hydrolysis was allowed to proceed for 0, 1, 2, and 3 min. The linear reaction rates were used to construct Lineweaver-Burk plots from which the kinetic constants were obtained. Any ^{32}P$_i$ found at zero time was subtracted from all other time points to correct the free ^{32}P$_i$ contamination in the ATP and BzATP solutions (less than 4%).

BzATP was about 250 times lower than for ATP, BzATP was commensurately hydrolyzed 250 times more slowly than the natural nucleotide substrate. The resulting k_{cat}/K_m (an estimate of the limiting second order rate constant for the hydrolytic reaction) was thus identical for both substrates, indicating that both are probably acted upon at the same kind of catalytic site on the BF_1, with the BzATP perhaps possessing a slower off-rate than ATP due to the hydrophobicity of the benzophenone group.

We measured the inhibition of ATPase activity by BzATP and BzADP, respectively, and found that both analogs behaved competitively, as shown in Figs. 1 and 2: K_i(BzATP) = 0.85 μM; K_i(BzADP) = 0.74 μM. Thus, both Bz-nucleotides interact non-covalently and with equivalent affinity at the catalytic site(s) of BF_1.

COVALENT PHOTOAFFINITY LABELING

Illumination of BF_1 with BzATP (<u>minus</u> added Mg^{2+} indicated a rapid and efficient inhibition of ATPase, which was pseudo-first order for more than one decade of decreasing enzyme activity. A proposed chemical mechanism for such

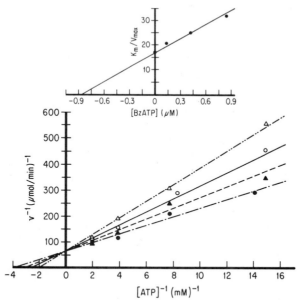

Fig. 1. Lineweaver-Burk Analysis and Slope Replot (inset) of BzATP Competitive Inhibition versus ATP Hydrolysis. The 1 ml reaction mixture contained: 50 mM Tris-acetate, pH 8.0, 1 mM $MgCl_2$, 250 ng F_1, and the indicated [γ-^{32}P]ATP concentrations, together with the following levels of BzATP: (●) no BzATP; (▲) 0.14 μM; (○) 0.42 μM; (△) 0.84 μM. AT Pase activity was measured as for Table I.

241

photocovalent ligand formation with BzATP is shown in Fig. 3, which illustrates how the light-generated triplet diradical might serve to covalently crosslink with the enzyme in a rapid, two-step, electron-pairing process (Williams et al., 1986). Such pseudo-first order photoinhibition of BF_1 by BzATP displayed rate-saturation kinetics over a range of photoprobe concentrations (Fig. 4), thereby supporting the following reaction pathway:

$$BF_1 + BzATP \underset{k_{-1}}{\overset{k_1}{\rightleftharpoons}} BF_1::BzATP \overset{k_{inact}}{\rightleftharpoons} BF_1\text{-BzATP}$$

where $k_{-1} + k_{inact} / k_1 = K_{app}$ (the apparent affinity of BzATP for BF_1 in the reversible step, under photolytic conditions); $K_{app} = 0.1 \ \mu M$. ATP afforded BF_1 excellent

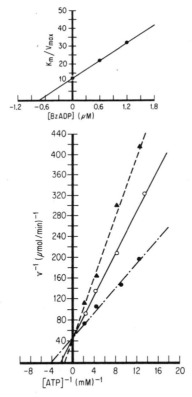

Fig. 2.   Lineweaver–Burk Analysis and Slope Replot (inset) of BzADP Competitive Inhibition versus ATP Hydrolysis.   The 1 ml reaction mixture contained: 50 mM Tris-acetate, pH 8.0, 1 mM $MgCl_2$, 250 ng F_1 and the indicated $[\gamma\text{-}^{32}P]ATP$ concentrations, together with the following levels of BzADP: ($\bullet$) no BzADP; ($\circ$) 0.6 μM BzADP; ($\blacktriangle$) 1.2 μM BzADP.   ATPase activity was measured as for Table I.

protection against photoinactivation by BzATP. These protection studies, also performed <u>minus</u> Mg^{2+}, yielded a $K_{protect}$ value (for ATP) of approx. 0.3 μM. The K protect value was derived from the x-intercept of Fig. 5, taking into account the competition between BzATP and ATP for a catalytic site, according to:

$$\text{"true" } K_{protect} = \text{x-intercept}/1 + ([BzATP]/K_{app})$$

For these experiments, the x-intercept (i.e., the "observed" $K_{protect}$) = 2.8 μM when [BzATP] = 0.84 μM. These data, together with results that showed an unchanged $K_m(ATP)$ when BF_1 activity was first photoinhibited by 50% with BzATP, suggest that the BzATP-induced photoinactivation is an all-or-none process characteristic of specific active site modification. It is important to emphasize that benzoylbenzoic acid, the photoreactive moiety of BzATP, was only able to inhibit BF_1 after prolonged illumination.

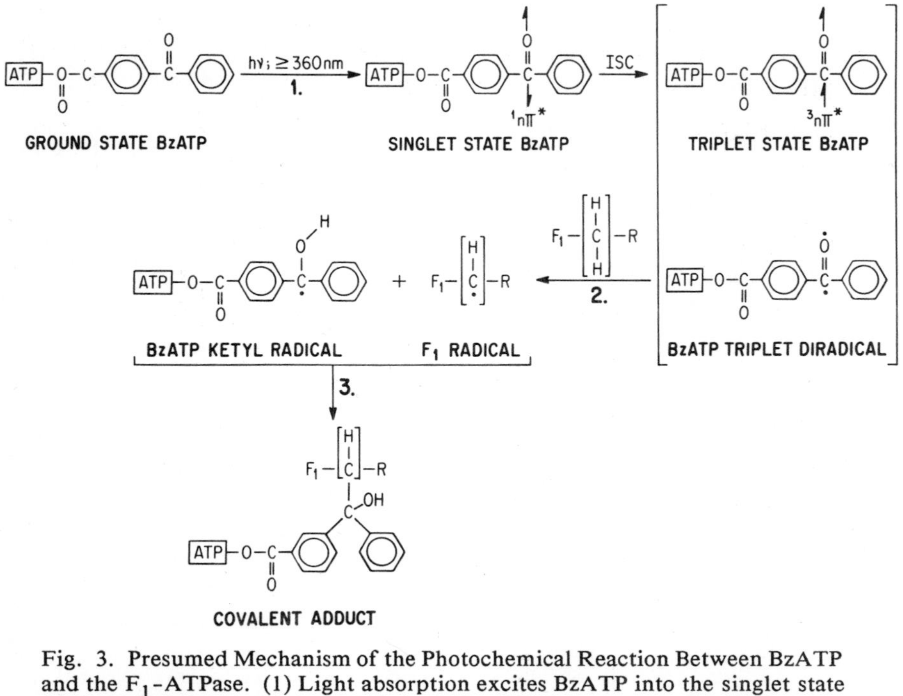

Fig. 3. Presumed Mechanism of the Photochemical Reaction Between BzATP and the F_1-ATPase. (1) Light absorption excites BzATP into the singlet state (paired electron spins), which is followed by a non-radiative process (intersystem crossing, ISC) to yield triplet state BzATP (unpaired electron spins). The reactive chemical intermediate of BzATP is proposed to be a triplet diradical, as indicated in the bracket. (2) Diradical BzATP abstracts a hydrogen atom from F_1 creating a radical pair. (3) Covalent bond formation between BzATP and F_1.

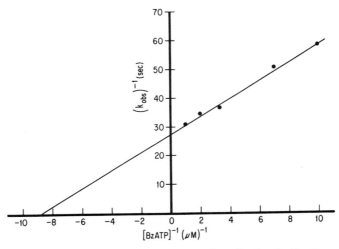

Fig. 4. Secondary Reciprocal Plot of F_1 Photoinactivation in the Presence of Increasing BzATP. The 1.5 ml photolysis incubation contained: 10 mM KH_2PO_4 pH 7.4, 13 μg/ml F_1, and BzATP as indicated. ATPase activity was measured spectrophotometrically in the presence of an ATP regenerating system (Pullman et al., 1960). The k_{obs} values were derived from semilog plots of the raw data.

Such slow inhibition could not be prevented by the presence of ATP, which indicated that the loss of enzyme activity in this latter case, unlike that observed with BzATP, was due to non-specific covalent modification of the enzyme.

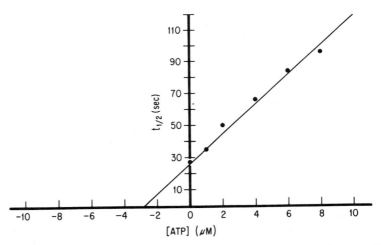

Fig. 5. Slope Replot of ATP Protection Against BzATP Photoinactivation of F_1 ATPase Activity. The 1.5 ml photolysis incubation contained: 10 mM KH_2PO_4 pH 7.4, 15 μg/ml F_1, and 0.84 μM BzATP in the absence and presence of several ATP concentrations. ATPase activity was measured as described for Fig. 4. The $T_{1/2}$ values were derived from semilog plots of the raw data.

IMPLICATIONS FOR COOPERATIVITY IN STUDIES ON THE LIMITING BINDING STOICHIOMETRY FOR COMPLETE PHOTOINHIBITION OF BF_1 BY BzATP

A procedure was devised for the complete resolution of $[^3H]BzATP$ covalently bound to BF_1 from tight, but non-covalently associated ^3H-labeled probe subsequent to actinic illumination. The method involved column separation of free from bound probe on Sephadex G-50, subsequent to photolysis of BF_1 in the presence of $[^3H]BzATP$, with an elution buffer containing 0.1% SDS. Utilizing this procedure, together with assays that yielded the extent of ATPase inhibition during photolysis with a 4- to 5-fold molar excess of probe:enzyme, we obtained the data of Fig. 6A. The experimental trials, indicated by open and closed circles, under conditions where a molar excess of BzATP was present, showed that two sites are capable of covalent modification, coincident with complete loss of ATPase activity. The solid curve in Fig. 6A was derived by fitting the data to the expression (Richardson and Ruteshouser, 1986; Wang, 1985):

$$y = (1-z)^n \times 100$$

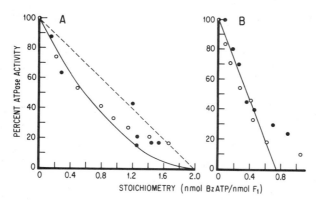

Fig. 6. Stoichiometry Plots for Covalently Bound BzATP Correlating with F_1 Photoinactivation. (A) The 1 ml photolysis incubation contained: 10 mM Na_2HPO_4 pH 7.4, 10.8 μM $[^3H]BzATP$, and either ($\bullet$) 2.7 μM F_1 or (O) 2.3 μM F_1. At various times (15s – 4 min), 1 μl aliquots were tested for ATPase activity and 50 μl aliquots were resolved into covalently bound and free $[^3H]BzATP$ chromatographically on Sephadex G-50 columns equilibrated with 10 mM KH_2PO_4, pH 7.4, that contained 0.1% SDS. (B) The 1 ml photolysis incubation for both experiments contained: 10 mM Na_2HPO_4 pH 7.4, 2.4 μM F_1, and initially, 2.7 μM $[^3H]BzATP$. In one experiment ($\bullet$), after the enzyme had been inactivated 30%, a second portion of $[^3H]BzATP$ (3.4 μM) was added and photolysis was continued to obtain 80% inactivation. In a second experiment (O), after the enzyme had been inactivated 50%, small portions of $[^3H]BzATP$ (approximately 0.7 μM) were added, prior to each remaining data point obtained, until 90% inactivation was achieved. Binding and activity assays were performed as described for (A).

where y = % remaining ATPase activity; n = <u>number</u> of assumed cooperatively interacting binding sites(or binding sites per cooperative catalytic unit)/mol BF_1; and z = <u>fraction</u> of the nucleotide binding sites/mol BF_1 that contain covalently bound [³H]BzATP. If n is taken as 2 or 3, the curves are nearly identical for our data under the conditions of these experiments. The principle underlying this analysis assumes that BF_1 may possess independently operating, adenine nucleotide catalytic sites (n = 1), or alternatively, 2 or 3 <u>cooperatively</u> interacting nucleotide sites (n = 2 or 3). If each nucleotide binding site behaved independently upon catalysis during photoinhibition by BzATP, Fig. 6A would yield a linear relationship between the % remaining activity and potential molar stoichiometries of 1, 2 or 3 moles of covalently bound probe/mol enzyme. Clearly, the non-linearity of the data in Fig. 6A do not satisfy these criteria. Rather, the curve obtained implies that cooperative interaction indeed occurs between at least 2 of the nucleotide binding domains on the enzyme. However, because the loss of activity <u>versus</u> stoichiometry of covalently bound probe/mol enzyme is curved and not linear (the dashed line), one cannot determine a <u>minimal</u> binding stoichiometry necessary for total inhibition of ATPase under these experimental conditions.

We therefore devised a more sensitive procedure for ascertaining this minimum binding stoichiometry coincident with total loss of activity. This approach consisted of the titration of sub-stoichiometric amounts of [³H]BzATP with BF_1 during photolysis, with aliquots removed during the time course of photolysis, and assayed for both remaining ATPase activity and for the covalent binding stoichiometry. Under these conditions, which minimized the concentration of free BzATP during photolysis, the data of Fig. 6B were generated. In these experiments (indicated by open and closed circles) we were able to titrate small amounts of probe and obtain a linear decrease in ATPase activity to about 85% inhibition. This strategy yielded a molar stoichiometry of nearly 1, correlated with complete enzyme inhibition. Thus, although there appear to be at least two catalytically competent nucleotide binding sites available for BzATP, the occupation of only one of them completely inhibits catalytic activity. This result supports the concept of cooperative nucleotide site (subunit?) interaction during the hydrolysis of ATP by BF_1, proposed by others (Grubmeyer et al., 1982; Cross et al., 1982; Penefsky, 1985a; Penefsky, 1985b; Gresser et al., 1982) using rather different experimental approaches.

WHICH SUBUNIT BINDS [³H]BzATP WITH UNIMOLAR STOICHIOMETRY?

It is noteworthy that the observation of 1 mol BzATP bound/mol BF_1 for complete inhibition of ATPase is corroborated by the pseudo-first order decay in activity achieved during photolysis (from data used to construct Fig. 4). Thus, following short-term photolysis incubations in the presence of sub-stoichiometric amounts of $[^3H]$BzATP, which yielded a 40% loss of enzyme activity (and generated a covalent stoichiometry = 0.3 mol probe/mol BF_1), we resolved the radiolabeled BFl subunits on 8% SDS/urea polyacrylamide gels (Fig. 7). The location of the ^3H-labeled subunit was determined via fluorographic exposure of the gel to X-ray film (Fig. 8), and indicates that under conditions of photolysis with sub-stoichiometric amounts of the probe relative to enzyme, the β subunit of BF_1 is labeled by the $[^3H]$BzATP almost exclusively.

The data of Figs. 6A and 6B, considered together with those of Fig. 8 provide a clear indication that BzATP, a photoaffinity substrate for the catalytic sites of BF_1, when present at levels equimolar to enzyme, interacts covalently with the enzyme at only one of at least two available catalytic loci.   Most importantly, our results demonstrate that the covalent occupancy by BzATP of this one nucleotide binding site, located on the β subunit, is sufficient to completely inactivate one mol of BF_1.   We

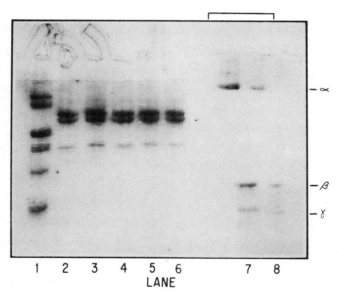

Fig. 7. Photograph of a Dried SDS–Polyacrylamide Gel of F_1 in Preparation for Fluorographic Analysis of the BzATP-Modified Subunit. The enzyme was partially, covalently modified with $[^3H]$BzATP, resolved from free radioactive ligand (as described for Fig. 6), and subsequently run on an 8% acrylamide gel (Kadenbach et al., 1983). Lane 1 contained molecular weight protein markers (Pharmacia). The amount of F_1 protein loaded per gel lane was: 4.8 μg (lanes 2, 4, 5, 6, 8), 8.6 μg (lane 3), and 9.5 μg (lane 7). The bracket shows that lanes 7 and 8 were excised and cut between the α and β bands. The location of the F_1 subunits in the dried gel are indicated. It is important to note that the placement of the excised α and β bands on the 3 MM paper, prior to drying the gel, resulted in the α and β bands no longer being correctly aligned with each other. Thus, this photograph serves as a reference for the location of the α and β subunit bands on the X-ray film image which follows.

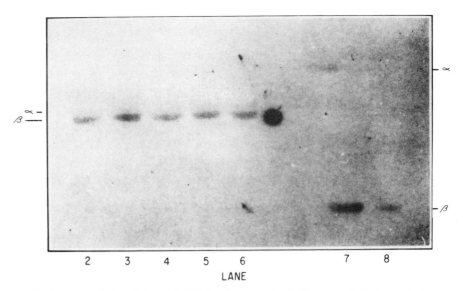

Fig. 8. X-ray Film of the Dried Gel Given in Fig. 7, Developed After 2 Weeks of Exposure to [³H]BzATP. The α and β bands are indicated and the β subunit location of covalently bound BzATP is clear.

conclude that a cooperative interaction occurs between at least two of these available nucleotide binding sites during catalysis. Apparently, the occupancy of one β subunit catalytic domain with nucleotide is an event that possesses information capable of being communicated to a second (and possibly a third) nucleotide-binding catalytic domain, a situation that defines a catalytic cooperativity mechanism for BF_1-ATPase.

REFERENCES

Bar-Zvi, D. and Shavit, N. (1984) Biochim. Biophys. Acta 765, 340-346.

Cross, R.L., Grubmeyer, C. and Penefsky, H.S. (1982) J. Biol. Chem. 257, 12101-12105.

Gresser, M.J., Myers, J.A. and Boyer, P.D. (1982) J. Biol. Chem. 257, 12030-12038.

Grubmeyer, C., Cross, R.I. and Penefsky, H.S. (1982) J. Biol. Chem. 257, 12092-12100.

Kadenbach, B., Jarausch, J., Hartmann, R. and Merle, P. (1983) Anal. Biochem. 129, 517-521.

Kambouris, N.G. and Hammes, G.G. (1985) Proc. Natl. Acad. Sci. USA 82, 1950-1953.

Mahmood, R. and Yount, R.G. (1984) J. Biol. Chem. 259, 12956-12959.

Manolson, M.F., Rea, P.A. and Poole, R.J. (1985) J. Biol. Chem. 260, 12273-12279.

Penefsky, H.S. (1979) Methods Enzymol. 55, 304-308.

Penefsky, H.S. (1985a) Proc. Natl. Acad. Sci. USA 82, 1589-1593.

Penefsky, H.S. (1985b) J. Biol. Chem. 260, 13735-13741.

Pullman, M.E., Penefsky, H.S., Datta, A. and Racker, E. (1960) J. Biol. Chem. 235, 3322-3329.

Richardson, J.P. and Ruteshouser, E.C. (1986) J. Mol. Biol. 189, 413-419.

Wang, J.H. (1985) J. Biol. Chem. 260, 1374-1377.

Williams, N. and Coleman, P.S. (1982) J. Biol. Chem. 257, 2834-2841.

Williams, N., Ackerman, S. and Coleman, P.S. (1986) Methods Enzymol. 126, 667-682.

ON THE NUMBER OF CATALYTIC SITES IN THE F_1-ATPase

THAT CATALYZE STEADY STATE ATP HYDROLYSIS

William S. Allison, John G. Verburg, and David A. Bullough

Department of Chemistry, M-001, University of California at
San Diego, La Jolla, CA 92093

The bovine heart mitochondrial F_1-ATPase (MF_1), which is composed of five different polypeptide chains in a subunit stoichiometry of $\alpha_3\beta_3\gamma\delta\epsilon$, has a total of six adenine nucleotide binding sites (Esch and Allison, 1979; Cross and Nalin, 1982). Kinetic evidence presented by Lardy's laboratory (Lardy et al., 1975; Schuster et al., 1975) suggested the presence of regulatory sites that specifically bind adenine nucleotides and catalytic sites which are less specific and hydrolyze ATP, ITP, GTP and other nucleoside triphosphates. Gautheron and her colleagues (DiPietro et al., 1980; Baubichon et al., 1981; Fellous et al., 1984) have provided evidence for a regulatory site on the β subunit of MF_1 which responds to ADP but not to GDP and GDP analogues.

We have recently shown that two structurally related affinity labels, 5'-fluorosulfonylbenzoyladenosine (FSBA) and 5'-p-fluorosulfonylbenzoylinosine (FSBI), modify different sites when they inactivate the bovine mitochondrial F_1-ATPase (Bullough and Allison, 1986a,b). Complete inactivation of the ATPase by FSBA proceeds with modification of either Tyr-368 or His-427, in mutually exclusive reactions which are pH-dependent, in all three copies of the β subunit. On the other hand, complete inactivation of the ATPase by FSBI proceeds with the modification of Tyr-345 in only a single copy of the β subunit. It was also observed that complete inactivation of ITP hydrolysis by FSBA occurred on modification of either Tyr-368 or His-427 in a single copy of the β subunit. Inactivation of ITPase activity by FSBI was accompanied by modification of Tyr-345 in a single copy of the β subunit.

Penefsky and his colleagues (Grubmeyer et al., 1982; Cross et al., 1982) have described the kinetic characteristics of ATP hydrolysis at a single catalytic site of MF_1. The essential features of single site catalysis include: (a) rapid, high affinity binding of

Molecular Structure, Function, and Assembly of the ATP Synthases
Edited by Sangkot Marzuki
Plenum Press, New York

substoichiometric ATP; (b) rapid equilibration of bound ATP with bound products giving an equilibrium constant of about 1; and (c) slow release of products unless excess ATP is added to bind to a second catalytic site. To provide evidence for or against our contention that FSBA modifies three regulatory sites and FSBI modifies a single catalytic site on complete inactivation of ATP hydrolysis, experiments were initiated to assess the capacity of MF_1 to catalyze ATP hydrolysis at a single catalytic site after inactivating steady state ATP hydrolysis with FSBA or FSBI. In the course of this investigation it became clear that the single high affinity catalytic site described by Grubmeyer et al. (1982) and Cross et al. (1982) does not participate directly in steady state hydrolysis of ATP by MF_1. The results that led to this conclusion are presented here.

RESULTS

Characteristics of Hydrolysis of Substoichiometric [α,γ-^{32}P]ATP by Unmodified MF_1

Fig. 1 compares the hydrolysis of substoichiometric [α,γ-^{32}P]ATP by native MF_1 with and without a prior incubation with 2 mM Pi. Penefsky and Grubmeyer (1984) have reported that it is essential to activate MF_1 with oxyanions to observe the characteristic features of single site catalysis. It is clear from Fig. 1 that our preparation of MF_1 exhibits these properties in the absence of activation with oxyanions. However, aging the enzyme for 1 h in the presence of 2 mM Pi before examining hydrolysis at a single catalytic site shifted the bound ADP/ATP ratio from about 2 to about 1. The second order rate constants determined in dilute solutions for the binding of labeled ATP to enzyme with and without aging with 2 mM Pi were 8.4 x 10^6 $M^{-1}s^{-1}$ and 7.0 x 10^6 $M^{-1}s^{-1}$, respectively. These values compare favorably with the rate constant of 6.0 x 10^6 $M^{-1}s^{-1}$ determined by Grubmeyer et al. (1982) for the binding of substoichiometric ATP to their preparation of MF_1 aged in the presence of Pi.

Hydrolysis of Substoichiometric [α,γ-^{32}P]ATP by MF_1 Modified by FSBA or FSBI

After inactivating the steady state activity of MF_1 by 98% with FSBA, the enzyme retained the essential features of single site catalysis described by Grubmeyer et al. (1982) as shown in Fig. 2. Although single site hydrolysis of ATP by enzyme modified with FSBA was carried out in the absence of Pi or aging with Pi, the apparent equilibrium constant for the hydrolysis of bound ATP was about 1, resembling single site ATP hydrolysis by unmodified MF_1, aged with Pi shown in Fig. 1B. The rate of binding of ATP to MF_1, modified with FSBA, in dilute solutions was 2.9 x 10^5 $M^{-1}s^{-1}$, which is about 20-fold less than the rate of binding of ATP to the unmodified enzyme.

The net rate of single site hydrolysis of ATP, which is limited by the dissociation of ADP, was determined to be 2.9 x 10^{-3} s^{-1}. This is the same rate observed when unmodified enzyme, with or without prior aging with 2 mM Pi, hydrolyzes ATP at a single catalytic site.

Fig. 2 also shows the profiles of ^{32}P labeled species that developed when a substoichiometric quantity of [α,γ-^{32}P]ATP was added to MF$_1$ in which the steady state activity had been attenuated by 94% by modification with FSBI. The rate of binding of substoichiometric ATP to enzyme modified with FSBI was 2.9 x 10^4 $M^{-1}s^{-1}$, which is about 200-fold less than the rate of binding of substoichiometric ATP to the native enzyme. The rate of release of [α-^{32}P]ADP from the single catalytic site of MF$_1$ modified with FSBI was 5.8 x 10^{-3} s^{-1}, which is nearly twice that determined for the unmodified enzyme or enzyme modified with FSBA.

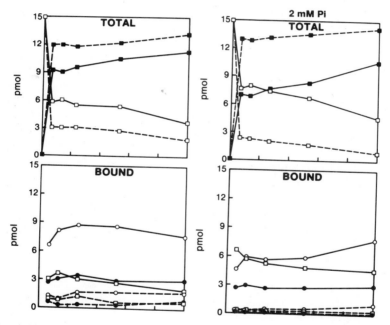

Fig. 1. Single Site Catalysis by MF$_1$ with and without Aging with 2 mM Pi. The reaction mixtures contained 0.3 μM [α,γ-^{32}P]ATP in 40 mM Tris-MES, pH 7.5, which contained 0.25 M sucrose, 0.5 mM MgSO$_4$ and also containing either: *Left*, 1 μM MF$_1$; or *Right*, 1 μM MF$_1$ which had been aged 1 h in the presence of 2 mM Pi. *Total*: reaction mixtures were quenched with perchloric acid and the radioactive species were determined by anion exchange HPLC. *Bound*: reaction mixtures were passed through Sephadex G-50 centrifuge columns before quenching with perchloric acid. The symbols used are: ADP, (O); Pi, (●); ATP, (□) and ADP/Pi, (■). The solid lines represent hydrolysis observed in the absence of promotion and the dashed lines represent hydrolysis after 10 s promotion by 5 mM ATP.

The observation that inactivation of the steady state activity of MF$_1$ with FSBI has a much greater effect on hydrolysis of ATP at a single catalytic site than does inactivation of the steady state activity of the enzyme with FSBA, supports our contention that FSBI modifies a single catalytic site and FSBA modifies regulatory sites (Bullough and Allison, 1986b).

Promotion of Hydrolysis of [α,γ-^{32}P]ATP at the Single Catalytic Site of MF$_1$ by ATP and ADP

In the course of this study, it was observed that 5 mM ADP is as effective as 5 mM ATP in promoting hydrolysis of [α,γ-^{32}P]ATP bound at a single catalytic site as well as in promoting release of radioactive products. This phenomenon was examined more closely by determining the concentration dependence of non-radioactive ATP or ADP added as promoter on hydrolysis of [α,γ-^{32}P]ATP bound at a single catalytic site of MF$_1$, with and without activation with 2 mM Pi. The results of these experiments are illustrated in Fig. 3. When added to final concentrations in the low micromolar

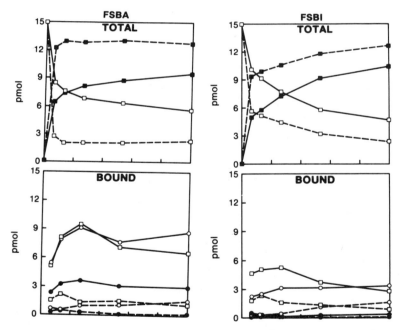

Fig. 2. Single Site Catalysis by MF$_1$ Modified with Either FSBA or FSBI. Reaction mixtures containing MF$_1$ inactivated by 98% with FSBA or 94% by FSBI were prepared as described in the legend of Fig. 1. Hydrolysis of substoichiometric [α,γ-^{32}P]ATP by *Left*, MF$_1$ modified with FSBA, and *Right*, MF$_1$ modified with FSBI was carried out as described in the legend of Fig. 1. The symbols used are: ADP, (O); Pi, (●); ATP, (□); and ADP/Pi, (■). The solid lines represent hydrolysis observed in the absence of promotion and the dashed lines represent hydrolysis after 10 s promotion by 5 mM ATP.

range, ADP is somewhat less effective than ATP with both forms of enzyme in promoting hydrolysis of $[\alpha,\gamma\text{-}^{32}P]$ATP bound at a single catalytic site.

From plots of the dependence of the estimated rate of hydrolysis of $[\alpha,\gamma\ ^{32}P]$ATP bound at a single catalytic site on promoter ATP concentration, the apparent second order rate constants of $9.4 \times 10^3\,M^{-1}s^{-1}$ and $2.6 \times 10^4\,M^{-1}s^{-1}$ were estimated for the apparent binding of ATP to promoter sites on MF_1 in the presence and absence of prior aging of the enzyme with 2 mM Pi, respectively. These values are at least 250-fold less than the second order rate constant of $6.3 \times 10^6\,M^{-1}s^{-1}$ determined by Cross et al. (1982) for the apparent binding of ATP to a second catalytic site during hydrolysis of a 20-fold excess of $[\gamma\text{-}^{32}P]$ATP by MF_1 aged with 2 mM Pi. The second order rate constants determined from the promotion experiments illustrated in Fig. 3 are comparable to the second order rate constant of about $10^3\,M^{-1}s^{-1}$ determined by Penefsky (1985) for promotion by ADP of hydrolysis of ATP bound to a single catalytic site of membrane-bound MF_1. Since Cross et al. (1984) have reported that the second order rate constant for the binding of ADP to a high affinity catalytic site is $10^5\,M^{-1}s^{-1}$, Penefsky (1985) has suggested that the apparent second order rate constant that he determined for the binding of ADP to a promotion site may reflect two rate constants, a rapid bimolecular binding step followed by a slower conformational change. A similar two step process might occur during the promoted hydrolysis of $[\alpha,\gamma\text{-}^{32}P]$ATP bound to a single catalytic site of MF_1 by low micromolar concentrations of ATP or ADP.

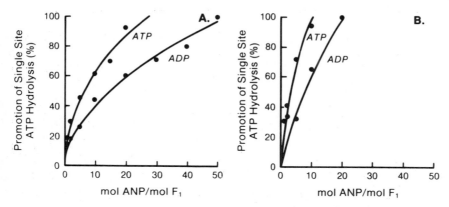

Fig. 3. Promotion by ATP or ADP of $[\alpha,\gamma\text{-}^{32}P]$ATP Hydrolysis and Product Release at a Single Catalytic Site of MF_1 with or without Aging with 2mM Pi. Reaction mixtures containing 1 μM modified or unmodified enzyme and 0.3 μM $[\alpha,\gamma\text{-}^{32}P]$ATP in 50 μl of 40 mM Tris-40 mM MES, pH 7.5, containing 0.25 M sucrose and 0.5 mM MgSO$_4$ were incubated for 1 min at 23°C, at which time 5.5 μl of ATP or ADP solutions were added to give the final concentrations indicated. After an additional 10 s, the reactions were quenched by the addition of 10 μl of 14% perchloric acid. The labeled species in the samples were analyzed by anion exchange HPLC. (A) MF_1 aged with 2 mM Pi; (B) MF_1 in the absence of aging with Pi.

Penefsky and Grubmeyer (1984) have reported that 1 nM MF_1 catalyzes the hydrolysis of 1 μM ATP at a rate of about 8 s^{-1}, which we have confirmed with our preparation of MF_1. It would thus be expected that low micromolar concentrations of ATP added as promoter should be hydrolyzed rapidly by 1 μM MF_1, the concentration of enzyme used in the promotion experiments. To confirm this, 5-50 μM [α,γ ^{32}P]ATP was added to 1 μM MF_1, preloaded with substoichiometric non-radioactive ATP, and incubated for 10 s before quenching the reaction with perchloric acid and analyzing the products as described in the legend to Fig. 3. This analysis showed that when 5 μM ATP was added as promoter, 93% was hydrolyzed in 10 s. In a separate experiment, only 45% of the preloaded [α,γ-^{32}P]ATP was hydrolyzed in 10 s. A similar discrepancy between the rate of ATP hydrolysis at the single site and promoter site was observed when 10, 15, and 20 μM ATP was used as promoter. When higher concentrations of promoter ATP were used, it was impossible to distinguish between hydrolysis at the pre-loaded single site and the promoter site without using quenched flow apparatus. However, from these experiments, it is clear that at promoter ATP concentrations of 5-20 μM, several turnovers of hydrolysis of promoter ATP occurred while hydrolysis of ATP at a preloaded catalytic site was incomplete.

DISCUSSION

The observation that the rate of promoted hydrolysis of [α,γ-^{32}P]ATP at a preloaded, single catalytic site is considerably slower than the rate of hydrolysis of micromolar concentrations of ATP added as promoter has important implications. This means that the single catalytic site characterized by Grubmeyer et al. (1982) and Cross et al. (1982) does not directly contribute to steady state ATP hydrolysis by MF_1 at these substrate concentrations. Therefore, it is not a "normal" catalytic site as proposed.

The results presented in this study can be accommodated by the model illustrated in Fig. 4. In this model, MF_1 is depicted as an asymmetric structure in which one of the $\alpha\beta$ pairs is tagged by the three minor subunits (Amzel et al., 1982; Pedersen and Amzel, 1985). Assuming that substoichiometric ATP binds to three catalytic sites of MF_1 randomly, one-third of the bound ATP will be on the catalytic site on the tagged β subunit and two-thirds of the bound ATP will be on the untagged β subunits. This is illustrated in Fig. 4, where, owing to space limitations, binding of substoichiometric ATP to only one of the untagged β subunits is shown. When promoter ATP (ANP) is added in micromolar concentrations, only one of the two vacant catalytic sites becomes occupied as shown in Scheme 2. Based on the results presented, we suggest that at these ATP concentrations, only one of the two catalytic sites hydrolyzes ATP at a rate commensurate with the turnover rate observed. To accommodate the observed slow hydrolysis of preloaded, substoichiometric ATP when a second catalytic site alone is occupied, we suggest that the single catalytic site characterized by Grubmeyer et al. (1982) and Cross et al. (1982) serves an activation role.

It is proposed in the model illustrated in Fig. 4 that all three catalytic sites are occupied when millimolar concentration of ATP are added to the enzyme, and that two of them, those on the untagged β subunits, catalyzed hydrolysis at the steady state rate, while the catalytic site on the tagged β subunit catalyzes hydrolysis at a much slower rate. This conclusion is based on observations reported by Cross et al. (1982). They observed, using quenched flow analysis, that the addition of 3.3 mM ATP to MF_1, preloaded with labeled, substoichiometric ATP, promoted the hydrolysis of about two-thirds of the labeled substrate at a rate commensurate with the steady state rate, while the promoted hydrolysis of the remaining one-third was considerably slower.

This model is compatible with the dependence on ATP concentration of the water-Pi oxygen exchange catalyzed during ATP hydrolysis by MF_1 reported by O'Neal

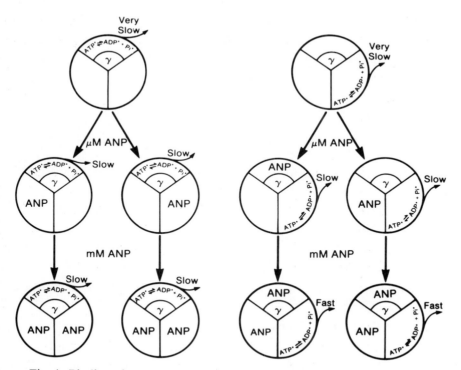

Fig. 4. Binding of Promoter ANP (ADP or ATP) to Vacant Catalytic Sites on MF_1 in which a Single, High Affinity Catalytic Site is Occupied with ATP.

and Boyer (1984). It is also consistent with the observation that the enzyme displays Km's of 1-2 μM and 160-250 μM with corresponding Vmax values of 2-4 μmol/min/mg and 50 μmol/min/mg, respectively, reported by Gresser et al. (1982) and Wong et al. (1984). We attribute the high affinity, slow catalysis observed to occur when two catalytic sites are occupied, with only one of them contributing to the steady state rate. The low affinity, fast catalysis is suggested to occur when three catalytic sites are occupied, with only two of them, those on untagged β subunits, contributing to the steady state rate.

REFERENCES

Amzel, L.M., McKinney, M., Narayanan, P. and Pedersen, P.L. (1982) Proc. Natl. Acad. Sci. USA 79, 5852-5856.

Baubichon, H., Godinot, C., DiPietro, A. and Gautheron, D.C. (1981) Biochem. Biophys. Res. Commun. 100, 1032-1038.

Bullough, D.A. and Allison, W.S. (1986a) J. Biol. Chem. 261, 5722-5730.

Bullough, D.A. and Allison, W.S. (1986b) J. Biol. Chem. 261, 14171-14177.

Cross, R.L. and Nalin, C.M. (1982) J. Biol. Chem. 257, 2874-2881.

Cross, R.L., Grubmeyer, C. and Penefsky, H.S. (1982) J. Biol. Chem. 257, 12101-12105.

Cross, R.L., Kironde, F.A.S. and Cunningham, D. (1984) in 3rd European Bioenergetics Conference, Vol. 3A, pp. 95-96, IUB-IUPAB Bioenergetics Groups, Hannover, West Germany.

DiPietro, A., Penin, F., Godinot, C. and Gautheron, D.C. (1980) Biochemistry 19, 5671-5681.

Esch, F.S. and Allison, W.S. (1979) J. Biol. Chem. 254, 10740-10746.

Fellous, G., Godinot, C., Baubichon, H., DiPietro, A. and Gautheron, D.C. (1984) Biochemistry 23, 5294-5299.

Gresser, M.J., Myers, J.A. and Boyer, P.D. (1982) J. Biol. Chem. 257, 12030-12038.

Grubmeyer, C., Cross, R.L. and Penefsky, H.S. (1982) J. Biol. Chem. 257, 12092-12100.

Lardy, H.A., Schuster, S.M. and Ebel, R.E. (1975) J. Supramol. Struct. 3, 214-221.

O'Neal, C.C. and Boyer, P.D. (1984) J. Biol. Chem. 259, 5761-5767.

Pedersen, P.L. and Amzel, L.M. (1985) in Achievements and Perspectives of Mitochondrial Research (Quagliarello, E., Slater, E.C., Palmieri, F., Saccone, C. and Kroon, A.M., eds.) Vol. 1, pp. 169-189, Elsevier Science Publishers, New York.

Penefsky, H.S. (1985) J. Biol. Chem. 260, 13728-13734.

Penefsky, H.S. and Grubmeyer, C. (1984) in H$^+$-ATPase (ATP Synthase) Structure, Function, Biogenesis of the F$_0$F$_1$ Complex of Coupling Membranes (Papa, S., Altendorf, K., Ernster, L. and Packer, L., eds.) pp. 195-204, ICSU Press, Adriatica Editrice, Bari, Italy.

Schuster, S.M., Ebel, R.E. and Lardy, H.A. (1975) J. Biol. Chem. 250, 7848-7853.

Wong, S-Y., Matsuno-Yagi, A. and Hatefi, Y. (1984) Biochemistry 24, 5004-5009.

THE EFFECTS OF NEUTRAL SALTS AND NUCLEOTIDES ON THE STABILITY

OF BEEF HEART MITOCHONDRIAL F_1-ATPase

Bruce Partridge and Sheldon M. Schuster

Department of Chemistry and School of Biological Sciences
University of Nebraska, Lincoln, Nebraska 68588-0304

INTRODUCTION

Analysis of upwards of 30 preparations of beef heart mitochondrial F_1-ATPase by size exclusion HPLC revealed that these highly purified enzyme preparations contain, on the average, 10 to 15% dissociated enzyme, or inactive polymers. In a few preparations, most notably several chloroform extracts (Spitsberg and Blair, 1977), as much as 50% of the total protein consists of subunits and high molecular weight aggregates. Despite the well known propensity of the major beef heart F_1 subunits[a] to form a white insoluble precipitate (e.g. Knowles and Penefsky, 1972), crystal clear solutions containing upwards of 2 mg/ml disaggregated subunits, as well as native F_1, may be obtained at room temperature. The cold lability of the enzyme is well known, and it might be surmised that dissociation occurs during long term storage at reduced temperatures. However no obvious correlation was found between the age of a particular F_1 preparation and the extent of dissociation.

Other studies, preparatory experimentation for nucleotide binding measurements, revealed that low mM concentrations of several divalent metal ions, notably Mn^{2+} and Ni^{2+} induce dissociation of the enzyme at room temperature. Under catalytic conditions, Mn^{2+} was also found to be a time-dependent inhibitor of ATPase activity. While the experimental implications of enzyme dissociation are obvious, and

---

[a]The abbreviations used are: F_1, Beef heart mitochondrial ATPase; SDS, sodium dodecyl sulphate; PAGE, polyacrylamide gel electrophoresis; HPLC, high pressure liquid chromatography.

the allosteric mechanism of F_1 implies a delicate balance of intersubunit forces, the factors which govern the stability of the enzyme at higher temperatures have never been thoroughly examined.

METHODS

F_1-ATPase was purified from beef heart mitochondria as by Knowles and Penefsky (1972) with the modification of Gruys et al. (1986). One mM freshly prepared phenylmethylsulfonyl fluoride was included in all buffers employed prior to the ion-exchange step in order to reduce protease contamination of the material (Partridge et al., in preparation).

Enzyme was desalted by two passes through P-6DG-packed (Bio-Rad) centrifugal columns equilibrated in 200 mM sucrose, 15 mM Na_2SO_4, 5 mM Tricine-NaOH, pH 8.0 (or as indicated). ATPase activity was determined at 30°C using a coupled NADH, lactate dehydrogenase, pyruvate kinase assay with 5 mM ATP, 6 mM $MgCl_2$ in 50 mM Tricine-NaOH, pH 8.0.

Measurement of protein dissociation: The dissociation of F_1 was monitored by size-exclusion HPLC using a Bio-Sil TSK-250 column (Bio-Rad, 300 x 7.5 mm) with 25 mM Pipes-NaOH, 100 mM Na_2SO_4, pH 6.5, and a flow rate of 1.2 ml/min. Samples (250 μl per injection) were injected via a 200 μl loop. Protein was detected by absorbance at 280 nm, and the peaks integrated to obtain concentrations. The buffers employed in the pH studies were Mes, pH 5.0 to 6.75; Mops, pH 6.5-7.75; Tricine, pH 7.5-8.75; Ches, pH 8.5-10.0. Buffers were prepared as 2 fold concentrates (100 mM) at the appropriate pH, including any salts, by titration with NaOH, KOH, LiOH, or NH_4OH depending on the cation salt added. For the experiments with divalent metal ions, the NaOH buffer was employed. Incubations at 30°C were initiated by mixture of equal, prewarmed volumes of F_1 (typically 0.5 mg/ml in 200 mM sucrose, 15 mM Na_2SO_4, 5 mM Tricine-NaOH, pH 8.0) and the salt containing buffer, so that carry over in all cases was 7.5 mM Na_2SO_4, 100 mM sucrose, 2.5 mM Tricine.

RESULTS

Fig. 1 depicts timecourses of the dissociation of F_1 at pH 8.0 and 30°C under several sets of conditions. The leftmost (void volume) peak is a soluble polymer which SDS-PAGE shows to contain all the F_1 subunits. The peak, which invariably increases with incubation time, has no detectable ATPase activity. The second peak represents catalytically active F_1, while the third and fourth peaks arise from the major and minor subunits.

When the enzyme is desalted and incubated in Tricine buffer (top panel) the F_1 peak attains a maximal height after 40 min incubation and thereafter decays through the original value at about 80 min. The ATPase activity of similarly desalted enzyme varies by as much as 40% over the first 30 min of incubation.

The inclusion of 200 mM sucrose in the incubation buffer (middle panel) results in a more rapid, and monotonic, decrease in the active peak. Despite the more rapid loss of the native enzyme, less of the soluble polymer is formed under these conditions. As described below, the occurrence of this aggregate is markedly sensitive to ionic

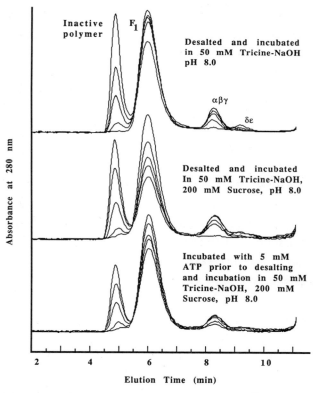

Fig. 1. Timecourses of Dissociation of F_1 at pH 8.0 and 30°C. F_1 was prepared in the indicated buffers and desalted by two passes through centrifugal columns. Following dilution to 0.25 mg/ml the samples were incubated at 30°C and analyzed by size exclusion HPLC at 1, 20, 40, 80, 120 and 200 min, top and middle panel; or 20, 60, 100, 140 and 220 min, bottom panel. The absorbance of F_1 at 280 nm was initially about 0.018.

effects. Addition of Na_2SO_4 (15 mM) to the sucrose containing incubation buffer slows dissociation to the extent that 85 to 90% of the enzyme remains intact over a five hour period.

Incubation of F_1 with 5 mM ATP for 10 minutes prior to desalting and incubation in Tricine-sucrose buffer (bottom panel) markedly stabilizes the enzyme. The effect is probably induced by the tight binding of ATP. Since HPLC analysis of F_1 storage buffer, and thoroughly desalted enzyme, shows a high proportion of AMP (not shown), there are certainly nucleotide binding sites for which ATP can effectively compete. Not unexpectedly, this incubation with ATP also minimizes the activity transients mentioned above. Direct inclusion of ATP (0.1 mM) in the incubation buffer is even more effective in stabilizing the enzyme. However, when incubation times exceed 1 h, a lag of up to several minutes duration becomes apparent at substrate concentrations below 0.5 mM ATP.

The effect of a variety of conditions on the stability of the enzyme at 30°C are depicted in Figs. 2 and 3. Since the dissociation of F_1 into subunits is apparently an irreversible process (Rosing et al., 1975; Penefsky and Warner, 1965), a single incubation time of one hour was employed to allow direct comparison of the rates of dissociation under the varying conditions.

At low salt concentrations the enzyme exhibits a broad region of stability between pH 6 and 8.5 (Fig. 2, top panel, no salt). The plot of stability is not symmetric about the optimal pH (7.5), and exhibits a shoulder between pH 6 and 7. Addition of 100 mM Na_2SO_4 narrows the region of stability by nearly half a pH unit on both sides of the optimum. $MgCl_2$ (2 mM) has nearly the same effect as 100 mM Na_2SO_4 from pH 5 to 8.5. Incubation with NaCl (100 mM) destabilizes the enzyme and induces a 20 to 25% dissociation of the enzyme in one hour even at pH 7.5. The effects of ammonium and sodium sulphates on the stability of the enzyme are indistinguishable. However the decrease in stability induced by 100 mM NH_4Cl is much less severe than with the corresponding sodium salt (not shown).

Nucleotides (Fig. 2, bottom panel), extend the pH range over which the enzyme is stable. Even at pH 5.0, little dissociation is apparent in the presence of 1 mM ATP. ITP and ADP are less stabilizing at low pH, but each of the nucleotides tested exaggerates the plateau of stability below pH 6.5. Thus the optimal pH shifts from 7.5 to 5.75, 6.75, and 5.75 in the presence of ATP, ADP, and ITP respectively. IDP (1 mM) or ADP (0.1 mM) have no stabilizing influence at pH 5.25 while 0.1 mM ATP does stabilize the enzyme. At the higher pH values, ADP is more stabilizing than ATP. At pH values greater than 7.0., ITP actually destabilizes the enzyme, relative to its behavior in low salt buffers.

Inclusion of 100 mM NaCl in the nucleotide containing buffers results in decreased stability at low pH. The stability of ATP incubated enzyme is unaffected by salt above pH 6.5, while NaCl completely obviates any stabilizing effect of ADP below pH 9.0.

Fig. 3 shows the effects of varying concentrations of monovalent cation containing salts on the stability of F_1 at pH 6 and 7.5. At the lower pH the destabilizing effect of anions generally followed the Hofmeister series $(SO_4 > NO_3 > Cl > I)$ at concentrations below 150 mM. However, contrary to the results of many studies of the effects of ions on protein salting in and salting out (Warren et

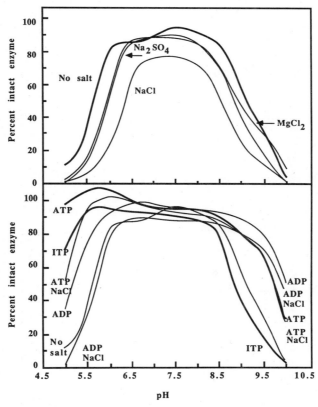

Fig. 2. Salt and Nucleotide Induced Dissociation of F_1 as a Function of pH. Enzyme (0.25 mg/ml) was incubated for 1 h at 30°C and analyzed by HPLC as described in Methods. The incubation buffers were: No salt, 50 mM buffer, 100 mM sucrose, with NaOH to the appropriate pH; Na_2SO_4, plus 100 mM Na_2SO_4; NaCl, 100 mM; $MgCl_2$, 2 mM; ATP, ADP, ITP 1 mM each; nucleotide plus NaCl, 1 mM and 100 mM respectively. The curves represent a spline fit of the 12 to 23 points for each experiment.

al., 1965; von Hippel et al., 1973; Hamabata and von Hippel, 1973; Hamabata et al., 1973; von Hippel, 1975), all the salts tested were found to destabilize the quaternary structure. The destabilization by monovalent cation sulphates follows no obvious pattern.

At pH 7.5, low concentrations of Na_2SO_4, K_2SO_4, $(NH_4)_2SO_4$, and Li_2SO_4 are all slightly stabilizing.

The effects of divalent metal ions on the stability of F_1 were also examined (Fig. 3) at pH 8.0. The divalents tested all catalyze the formation of insoluble precipitates (while monovalents favor free subunits or the soluble polymer) which makes quantitation of their effects more difficult. Small changes in the concentration of Ni^{2+} or Co^{2+} produce large changes in the dissociation of the enzyme. While a 1 h incubation with 12 $NiSO_4$ does not induce dissociation, the enzyme is 50% dissociated by 15 mM, while dissociation is complete within 20 min at 20 mM $NiSO_4$. Conversely,

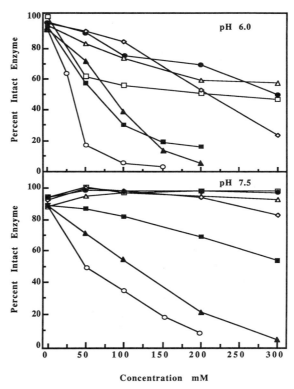

Fig. 3. Salt Dependence of F_1 Dissociation. F_1 (0.25 mg/ml) was incubated with various salts for 1 h at 30^{0}C in 100 mM sucrose, 50 mM Mes, pH 6.0 (top panel), or sucrose, 50 mM Tricine pH 7.5 prior to analysis by HPLC. Each buffer was prepared with the hydroxide of the appropriate cation: Na_2SO_4 □; $(NH_4)_2SO_4$ ●; K_2SO_4 △; Li_2SO_4 ◇; NaCl ■; NaI O; $NaNO_3$ ▲.

Ca^{2+} destabilizes the enzyme somewhat at 10 mM while 100 mM or more is required to induce 50% dissociation in 1 h. The ability of the various ions to catalyze ATP hydrolysis is shown for comparative purposes.

DISCUSSION

These experiments show that F_1 is extremely sensitive to the effects of neutral salts as compared to other enzymes (Warren et al., 1965). There is also some evidence of instability of the enzyme to the mechanical forces involved in centrifugal filtration (see Fig. 1, panel 1). While the later effects appear to be reversible, there is no evidence that the beef heart enzyme can be reassembled from subunits in vitro. The experimental implications of this instability are manifold, ranging from artifacts in nucleotide or chemical modifier binding experiments, to transients under the high stress conditions of stop flow. The biological significance of the instability is far less certain, but the apparent irreversibility of the dissociation does point to major differences between the beef heart and the bacterial ATPases, which freely dissociate and reassociate in solution (Hsu et al., 1984; Futai, 1977). Given the conditions of pH and salt concentration of the mitochondrium, it would not be surprising if the enzyme were unstable in its native environment.

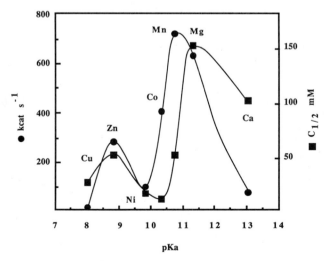

Fig. 4. Comparison of Catalytic and Dissociative Properties of Divalent Metal Ions. For kcat measurements see Urbauer et al., 1987. $C_{1/2}$ represents the concentration of divalent metal ion sufficient to induce 50% dissociation of a solution of F_1 (pH 8.0) in one hour. Analysis was by HPLC.

ACKNOWLEDGMENT

This work was supported by Grant PCM-09287 from the National Science Foundation.

REFERENCES

Gruys, K.J., Urbauer, J.L., Schuster, S.M. (1985) J. Biol. Chem. 260, 6533-6540.

Futai, M. (1977) Biochem. Biophys. Res. Comm. 79, 1231-1237.

Hamabata, A., Chang, S. and von Hippel, P.H. (1973) Biochemistry 12, 1271-1282.

Hamabata, A. and von Hippel, P.H. (1973) Biochemistry 12, 1264-1271.

Hsu, S-Y, Senda, M., Kanazawa, H., Tsuchiya, T. and Futai, M. (1984) Biochemistry 23, 988-993.

Knowles, A.F. and Penefsky, H.S. (1972) J. Biol. Chem. 247, 6617-6623.

Penefsky, H.S. and Warner, R.C. (1965) J. Biol. Chem. 240, 4694-4702.

Rosing, J., Harris, D.A., Kemp, A.Jr. and Slater, E.C. (1975) Biochim. Biophys. Acta 376, 13-26.

Spitsberg, V.L. and Blair, J.E. (1977) Biochim. Biophys. Acta 460, 136-141.

von Hippel, P.H. (1975) in *Protein-Ligand Interactions* (Sund, S. and Blauer, G., eds.) pp. 452-471, Walter de Gruyter, Berlin.

von Hippel, P.H., Peticolas, V., Schack, L. and Karlson, L. (1973) Biochemistry 12, 1256-1263.

Warren, J.C., Stowring, L. and Morales, M.F. (1965) J. Biol. Chem. 241, 309-316.

CHARACTERIZATION OF NUCLEOTIDE BINDING SITES ON CHLOROPLAST COUPLING FACTOR 1 (CF$_1$): EFFECTS OF NUCLEOTIDES ON NUCLEOTIDE TRIPHOSPHATE FORMATION BY ISOLATED CF$_1$

Hidehiro Sakurai and Toru Hisabori

Department of Biology, School of Education, Waseda University
Nishiwaseda 1, Shinjuku, Tokyo 160, Japan

ABSTRACT

CF$_1$ has several kinds of nucleotide binding sites. From our studies of nucleotide binding by UV difference spectroscopy and by equilibrium dialysis and of effects of methanol on the enzyme-bound ATP formation from medium ADP, we have characterized five kinds of nucleotide binding sites: Sites A, B, C and I, and a few low-affinity sites. Our conclusion is "CF$_1$ has two kinds of nucleotide tightly binding sites, and most of the tightly bound ADP is on the non-catalytic site".

MATERIALS and METHODS

CF$_1$ purified from spinach chloroplasts was deprived of dissociable nucleotides by slowly passing through a Sephadex G-25 column or repeatedly passing through centrifuge columns (G-50 fine) equilibrated with EDTA. The CF$_1$ thus prepared contained 1.0-1.4 mol adenine nucleotides per mol (Hisabori and Sakurai, 1985). Enhancement of CF$_1$-bound ATP formation by organic solvents and determination of ATP by luciferase assay were as described previously (Sakurai and Hisabori, 1987).

RESULTS and DISCUSSIONS

Isolated CF$_1$ retains about 1 mol of tightly bound ADP per mol of enzyme. Because the enzyme-bound ATP formation by isolated CF$_1$ did not depend on medium ADP, Feldman and Sigman (1982) concluded that the tightly-bound ADP was the substrate for ATP formation. More recently, Feldman and Boyer (1985) stated that

"Several lines of evidence indicate that some or most of the tightly bound ADP retained on spinach CF_1 (about 1 mol of ADP per mol of CF_1) is bound at an active site". On the other hand, we have reported that isolated CF_1 did not exchange the tightly bound ADP with medium ADP even in the presence of 20% methanol which elicits CF_1 Mg-ATPase activity, and concluded that the tightly bound ADP is not at the catalytic site (Hisabori and Sakurai, 1985).

We studied the effects of medium ADP on CF_1-bound ATP formation, and confirmed that it did not significantly increase the ATP formation in aqueous media. However, in the presence of methanol, medium ADP significantly increased the ATP formation (Sakurai and Hisabori, 1987). We have characterized five kinds of nucleotide binding sites on CF_1 (Fig. 1), and the properties of these sites will be described below:

Site A: tightly binding and non-catalytic site. When CF_1 was depleted of dissociable nucleotides by slowly passing through a Sephadex G-25 column, it still retained about 1 mol ADP per mol of CF_1. When the binding of labeled ADP was studied by equilibrium dialysis, we found that the total bound ADP was about 1 mol larger than the bound labeled ADP. As this ADP did not exchange with medium $[^{14}C]ADP$ even in the presence of 20% methanol which enhances Mg-ATPase of CF_1, we concluded that this binding site (Site A) is not a catalytic one (Hisabori and Sakurai, 1985). Gel-permeation chromatography in the presence of 50% glycerol could not remove ADP bound to this site.

Sites B and C: exchangeable and high-affinity sites with different affinities. These sites can bind not only ADP, but also CDP, GDP, IDP and UDP from aqueous

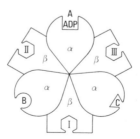

Fig. 1. A Model for the Nucleotide Binding Sites on CF_1. Site A: tightly binding and non-catalytic site, from which bound ADP is not removable. Site I: tightly binding and catalytic site, from which ADP is slowly removable by gel-permeation chromatography. Bound ADP can be converted to CF_1-bound ATP upon addition of Pi. Sites B and C: Exchangeable, noncatalytic and high-affinity sites with different Kds. These sites are responsible for the UV absorption changes of bound nucleotides and the hysteretic behaviors of CF_1. Sites II and III: low-affinity and exchangeable sites.

media, and the binding of these nucleotides induces UV absorption changes (Tanaka and Sakurai, 1980; Chiba et al., 1981; Hisabori and Sakurai, 1984). These spectral changes were not ascribable to the tyrosine spectrum as proposed by Penefsky (1979), but to the red shifts of nucleotide absorption bands accompanied by some decrease of absorbance (Fig. 2). More recently, similar studies with isolated TF_1 and its subunits from a thermophilic bacterium suggested that the cause of the red shifts could be an incorporation of nucleotide base moiety into hydrophobic environments of each subunits (Hisabori et al., 1986). ATP can also be bound to these sites.

The affinities of these sites (Sites B and C) for ADP were estimated to be 0.021 μM and 1.6 μM, respectively. The presence of EDTA decreased the affinities of ADP to these sites, but not the number of the sites. Ca^{2+} was as effective as Mg^{2+} in increasing the affinities of ADP to these sites. Girault et al. (1973) studied the interaction of adenine nucleotides with CF_1, using CD spectroscopy, and found that Pi, PPi or tripolyphosphate inhibited the interaction. We found that PPi which is not a competitive inhibitor of ATPase (Shinohara and Sakurai, 1982) strongly inhibited the UV spectral changes. High concentrations of Pi (a few mM range) also inhibited the UV spectral changes. The half time of the ADP binding was a few minutes (Tanaka and Sakurai, 1980) and 20% methanol only slightly accelerated the binding rate (Hisabori and Sakurai, 1984). From the above studies, we concluded that these high-affinity sites are not catalytic ones. The binding of nucleotides to sites B and C increased the resistivities of CF_1 to inactivation by high concentrations of methanol or by cold-anion treatment. Recent studies indicate that the binding of ADP to these sites are also responsible for the hysteretic behaviors of Mg-ATPase of isolated CF_1 (Hisabori et al., in preparation).

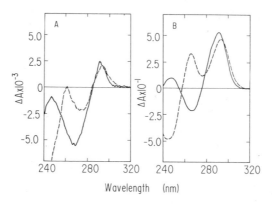

Fig. 2. (A) Difference Spectra Obtained by the Addition of 8 μM CDP (Solid Line) or 40 μM GDP (Dashed Line) to 2.3 μM CF_1, and (B) Those Obtained by Shifting the Nucleotide (20 μM Each) Absorption Spectra 10 nm to Longer Wavelengths.

Site I: catalytic site, on which ADP is somewhat tightly bound, but from which slowly removable in the presence of EDTA. Feldman and Sigman (1982) found that isolated CF_1 can synthesize enzyme-bound ADP and that medium ADP was ineffective as a substrate. These results seem to suggest that the substrate ADP is the so-called "tightly bound ADP". We found that although CF_1 prepared by passage through a single centrifuge column could synthesize a considerable amount of enzyme-bound ATP without addition of medium ADP, the amount of ATP synthesized was gradually decreased on repeatedly passing CF_1 through Sephadex G-50 centrifuge columns equilibrated with EDTA. The latter CF_1 preparation which still contained a slightly more than one ADP per enzyme could synthesize only a small amount of ATP without medium ADP. However, it could synthesize a large amount of enzyme-bound ATP when ADP was added in the presence of methanol (Sakurai and Hisabori, 1987) (Fig. 3). These results indicate that after extensive deprivation of ADP by repeated gel-permeation chromatography, only a small part of the tightly bound ADP remains on the catalytic site (Site I) and a large part of it is on the non-catalytic site (Site A).

Medium ADP is accessible to Site I only when CF_1 is properly activated (e.g. by methanol, ethanol or acetone) (Sakurai and Hisabori, 1987). Dimethyl sulfoxide which significantly enhances bound ATP formation in beef heart F_1 (Sakamoto et al., 1983), did not enhance the CF_1-bound ATP formation from medium ADP. Dimethyl sulfoxide did not enhance Mg-ATPase activity of isolated CF_1, either. Recent studies indicate that IDP can be a substrate for the CF_1-bound ITP formation.

Sites II and III: Equilibrium dialysis indicated that CF_1 has, in addition to the above four binding sites (Sites A, B, C and I), a few (probably two) low-affinity sites (Hisabori and Sakurai, 1984). The roles of these sites remain to be investigated.

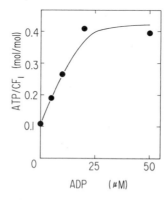

Fig. 3. Effect of the ADP Concentration on CF_1-Bound ATP Formation in the Presence of Methanol. The reaction mixture contained 0.75 μM CF_1, 25 mM MES (pH 6.0), 50 mM Pi (pH 6.0), 10 mM $MgCl_2$, 25% methanol and ADP as indicated. The reaction time was 30 min at 27°C.

ACKNOWLEDGMENT

This work was supported in part by Grant-in-Aid for Scientific Research on Priority Areas of "Bioenergetics" from the Ministry of Education, Science and Culture, Japan.

REFERENCES

Chiba, T., Suzuki, H., Hisabori, T. and Sakurai, H. (1981) Plant Cell Physiol. 22, 551-560.

Feldman, R.I. and Sigman, D.S. (1982) J. Biol. Chem. 257, 1678-1683.

Feldman, R.I. and Boyer, P.D. (1985) J. Biol. Chem. 260, 13088-13094.

Girault, F., Galmiche, J.-M., Michel-Villaz, M. and Thiery, J. (1973) Eur. J. Biochem. 128, 405-411.

Hisabori, T. and Sakurai, H. (1984) Plant Cell Physiol. 25, 483-493.

Hisabori, T. and Sakurai, H. (1985) Plant Cell Physiol. 26, 505-514.

Hisabori, T., Yoshida, M. and Sakurai, H. (1986) J. Biochem. 100, 663-670.

Penefsky, H.S. (1979) Adv. Enzymol. 49, 223-280.

Sakamoto, J. and Tonomura, Y. (1983) J. Biochem. 93, 1601-1614.

Sakurai, H. and Hisabori, T. (1987) in *Progress in Photosynthesis Research III* (Biggins, J., ed.) pp. 1.13-16, Martinus Nijhoff, Dordrecht.

Sakurai, H., Shinohara, K., Hisabori, T. and Shinohara, K. (1981) J. Biochem. 90, 95-102.

Shinohara, K. and Sakurai, H. (1981) Plant Cell Physiol. 22, 1447-1457.

Tanaka, M. and Sakurai, H. (1980) Plant Cell Physiol. 21, 1585-1591.

A HYDROPHOBIC PROTEIN, CHARGERIN II, PURIFIED FROM

RAT LIVER MITOCHONDRIA IS ENCODED IN THE UNIDENTIFIED

READING FRAME A6L OF MITOCHONDRIAL DNA

Tomihiko Higuti*, Tsutomu Negama*, Michio Takigawa*, Junji
Uchida*, Takeshi Yamane*, Tuyoshi Asai*, Isamu Tani*, Kenji Oeda**,
Masatoshi Shimizu**, Keiko Nakamura** and Hideo Ohkawa**

*Faculty of Pharmaceutical Sciences, University of Tokushima
Shomachi, Tokushima, 770, Japan and **Biotechnology Laboratory
Takarazuka Research Center, Sumitomo Chemical Co., Ltd.
Takatsukasa, Takarazuka, Hyogo, 665, Japan

SUMMARY

Previous studies showed that a hydrophobic protein called chargerin II may
have a key role in energy transduction of oxidative phosphorylation, since antibody
against chargerin II labeled with monoazide ethidium inhibited ATP synthesis, ATP-Pi
exchange, and reversed electron flow from succinate to NAD coupled with succinate
oxidation by O_2. In the present work, unlabeled chargerin II was purified from intact
rat liver mitochondria by high performance liquid chromatography on preparative gel
permeation and reverse-phase columns. Chargerin II was identified with antibody
against chargerin II labeled with monoazide ethidium and $[^{125}I]$-protein A. The
purified preparation of chargerin II, which was a single protein as judged by
polyacrylamide gel electrophoresis and Westernblotting, was digested with
lysylendopeptidase. The digest was separated on a reverse-phase column into five
peptides, which all crossreacted with the antibody against chargerin II labeled with
monoazide ethidium, indicating that they were fragments of chargerin II. The
sequences of two of these peptides (a total of 12 amino acids) were determined and
found to be highly homologous with the sequence of the carboxyl terminal peptide of
the putative polypeptide encoded by the unidentified reading frame A6L of
mammalian mitochondrial DNA, which overlaps the ATPase 6 gene. Thus we
concluded that chargerin II is encoded by this unidentified reading frame A6L gene of
the mitochondrial DNA.

INTRODUCTION

ATP synthase is a multi-subunit complex that utilizes a transmembrane proton gradient to form ATP (Mitchell, 1961). ATP synthase is composed of two domains: a hydrophilic part called F_1, which is the catalytic site of ATP synthesis and a membranous domain called F_0, is responsible for the energy transduction (Hatefi, 1985; Kagawa, 1984; Amzel and Pedersen, 1983; Kanazawa and Futai, 1982). The catalytic mechanism of F_1 may be the same in mitochondria, chloroplasts, yeasts, and bacteria, since the compositions and the primary structures of the subunits of their F_1s are similar and highly homologous. However, the compositions of the subunits of F_0 in prokaryotes and eukaryotes are quite different: in prokaryotes, F_0 has only three different subunits (Kanazawa and Futai, 1982) whereas in eukaryotes, it seems to consist of 7-13 different subunits (Hatefi, 1985). Little is known about the role in energy transduction of the additional subunits in F_0 of eukaryotes that are not present in F_0 of prokaryotes.

Recently we reported two hydrophobic proteins, named chargerin I and II (apparent molecular masses, 8 and 13 kDa), that bind anisotropic inhibitors (Higuti, 1984) of energy transduction in oxidative phosphorylation in an energy-dependent fashion in rat liver mitochondria (Higuti et al., 1985; Higuti et al., 1981). Antibody against chargerin II inhibited ATP synthesis, ATP-Pi exchange, and reversed electron flow from succinate to NAD coupled with succinate oxidation by O_2 in mitoplasts (inner membrane plus matrix) prepared from rat liver mitochondria (Higuti et al., 1985; Higuti et al., 1987b). Thus it seemed that chargerin II has an essential role in energy transduction in oxidative phosphorylation in rat liver mitochondria. In the present work we purified unlabeled chargerin II, which crossreacts with the antibody against chargerin II labeled with monoazide ethidium, from rat liver mitochondria by HPLC. Then we hydrolyzed the purified protein and found that the primary structures of its peptide fragments were highly homologous with the sequence of the carboxyl terminal peptide of the putative polypeptide encoded in the unidentified reading frame A6L (URFA6L) of mammalian mitochondrial DNA (Anderson et al., 1981; Grosskopf and Feldmann, 1981; Bibb et al., 1981; Anderson et al., 1982; Pepe et al., 1983). The URFA6L product is suggested to be associated with F_0 of mitochondrial ATP synthase (Macreadie et al., 1982; Linnane et al., 1985; Velours and Guerin, 1986; Fearnley and Walker, 1986).

MATERIALS AND METHODS

Rat (Wistar strain) liver mitochondria were isolated by the method of Soper and Pedersen (1979) and frozen at -20°C. Antiserum against chargerin II labeled with monoazide ethidium was prepared as described previously (Higuti et al., 1985).

274

Chargerin II was identified radioimmunochemically with the antibody against chargerin II labeled with monoazide ethidium and [^{125}I]-protein A, essentially by the method of Towbin et al. (1979) and Burnette (1981) as described in Higuti et al. (1987a). Other methods were as described (Higuti et al., 1987a).

RESULTS

Purification of Chargerin II by HPLC

The mitochondria were treated with a mixture of chloroform/methanol, 2:1 (vol/vol). Unlabeled chargerin II was extracted into the organic phase. After silica gel chromatography of the extract, fractions of chargerin II were applied to a column (21.5 x 600 mm) of TSK G3000SW that had been equilibrated with buffer consisting of 0.1% sodium dodecylsulfate, 0.02% sodium azide, and 100 mM sodium phosphate buffer at pH 7.0. Fractions of chargerin II were collected, dialyzed against deionized water, and concentrated in an Amicon, model 402, Diaflo Cell. Then, chargerin II obtained thus was applied to a reverse-phase column (20 x 250 mm) of Cosmosil 5C$_8$300 that had been equilibrated with 0.1% trifluoroacetic acid. Elution was carried out with a linear gradient of 0-80% solution of 94% isopropylalcohol, 5% acetonitrile and 0.1% trifluoroacetic acid. Fractions containing chargerin II were collected and concentrated in a Tomy Seiko Co., model CC-180, centrifugal concentrator. The residue obtained was solubilized in a solution of 3.3% sodium dodecylsulfate and 2% mercaptoethanol and dialyzed against deionized water. The resulting fraction of chargerin II was purified further by rechromatography on the same column with a lower gradient.

The purity of the preparation of chargerin II was increased efficiently by the present purification procedure and that the final preparation of chargerin II was highly purified as judged by sodium dodecylsulfate/urea-gel electrophoresis on linear 9-20% polyacrylamide gel and Western blotting (Higuti et al., 1987a).

Sequencing of Peptides of Chargerin II

First the sequence of the purified intact chargerin II (25 µg) was examined with an Applied Biosystems, model 470A, gas-phase sequenator, but no phenylthiohydantoin (PTH)-amino acid was detected. This could be due to blocking of the amino-terminal of the protein. So next the intact chargerin II was digested with lysylendopeptidase in the presence of 5 M urea and the digest was applied to a reverse-phase column. Five peaks of fragment peptides were obtained. All these peptides crossreacted with the antibody against chargerin II labeled with monoazide ethidium, indicating that they were fragments of chargerin II. The sequences of these peptides were analyzed with a gas-phase sequenator and the peptides in peaks 1 and 3 were found to have the sequences, Trp-Thr-Lys and Ile-Tyr-Leu-Pro-Leu-Ser-Leu-Pro-Pro, respectively (Higuti et al., 1987a).

Homology search in the data bank of the National Biomedical Research Foundation (NBRF) showed that the sequences of the fragment peptides of chargerin II determined in the present work were highly homologous with the sequence of the carboxyl terminal peptide of the putative polypeptide encoded in the unidentified reading frame A6L (URFA6L) of mammalian mitochondrial DNA, with identities of 13/13 (Wistar strain rats) (Pepe et al., 1983), 12/13 (Sprague-Dawley strain rats) (Grosskopf and Feldmann, 1981), 11/13 (mouse) (Bibb et al., 1981), 11/13 (bovine) (Anderson et al., 1982), and 9/13 (human) (Anderson et al., 1981) amino acids, as shown in Fig. 1. The present results are also consistent with the fact that lysylendopeptidase splits peptides on the carboxyl side of lysine residues (Masaki et al., 1981).

DISCUSSION

The primary structures of fragment peptides of chargerin II determined in the present work were highly homologous with the sequence of the carboxyl terminal peptide of the putative polypeptide encoded by the unidentified reading frame A6L (URFA6L) of mammalian mitochondrial DNA (Fig. 1). Thus, it seems very llikely that chargerin II purified in the present work is encoded by the URFA6L of mitochondrial DNA. It is noteworthy that the preparation of chargerin II obtained previously by a different method was a mixture of hydrophobic proteins labeled with monoazide ethidium (Higuti et al., 1985). But fortunately the antibody obtained was raised against only one sort of hydrophobic protein of rat liver mitochondria as shown in the present work and the previous paper (Higuti et al., 1987b).

The URFA6L product seems to be associated with F_0 of mitochondrial ATP synthase judging from the following observations: URFA6L overlaps the ATPase 6 gene (Anderson et al., 1981; Grosskopf and Feldmann, 1981; Bibb et al., 1981; Anderson et al., 1982; Pepe et al., 1983) and the same messenger RNA is used for translation of both these genes (cf. Chomyn et al., 1983), mutations in the yeast URFA6L affect the assembly or function of the ATP synthase (Macreadie et al., 1982; Linnane et al., 1985), the URFA6L product was found in ATP synthase (Velours and Guerrin, 1986; Fearnley and Walker, 1986) and anti-chargerin II antibody inhibited ATP synthesis and ATP-Pi exchange reactions in rat liver mitoplasts (Higuti et al., 1985; Higuti et al., 1987b).

Main common features of the URFA6L products and related proteins are as follows. (1) Sequence homology is very weak; only the first three amino acids, MPQ, are invariant. (2) Hydrophobic profiles suggest a single alpha-helical membrane-spanning domain. (3) There is a cluster of positive charges (Fig. 1).

Fig. 1 also shows that the partial sequence of chargerin II determined in the present work is highly conserved in mammals but is defected in *Xenopus* (Wong et al., 1983), the fruit fly (Wolstenholme and Clary, 1985), yeast (Macreadie et al., 1983; Velours et al., 1984), and *Aspergillus nidulans* (Grisi et al., 1982). Furthermore, no homologous protein has been found in prokaryotes. Therefore, it seems that during evolution, the URFA6L gene first appeared in a primitive eukaryote. It is uncertain whether as a result of evolutional change of the URFA6L gene, ATP synthase acquired a special device for more efficient energy transduction than that in prokaryotes. But, in relation to this, it is interesting that the URFA6L product binds 2.7 molecules of phosphate per molecule of protein (Blondin, 1979). It seems most unlikely that this phenomenon reflects a role in binding one of the substrates for the ATP synthase reaction or a role in transport of phosphate across the inner mitochondrial membrane, which were proposed previously, since phosphate as substrate of ATP synthesis is bound to F_1 (Lauquin et al., 1980) and a phosphate carrier protein is known (Wohlrab, 1980; Tommasino et al., 1987).

```
Chargerin II                                                          WTKIYLPLSLPP
RAT(Wistar)      MPQLDTSTWFITIISSMATLFILFQLKISSQTFPAPPSPKTMATEKTNNPWESKWTKIYLPLSLPPQ
RAT(S. Dawley)   MPQLDTSTWFITIISSMATLFNLFQLKISSQTFPAPPSPKTMATEKTNNPWESKWTKIYFPLSLPPQ
MOUSE            MPQLDTSTWFITIISSMITLFILFQLKVSSQTFPLAPSPKSLTTMKVKTPWELKWTKIYLPHSLPQQ
BOVINE           MPQLDTSTWLTMILSMFLTLFIIFQLKVSKHNFYHNPELTPTKMLKQNTPWETKWTKIYLPLLLPL
HUMAN            MPQLNTTVWPTMITPMLLTLFLITQLKMLNTNYHLPPSPKPMKMKNYNKPWEPKWTKICSLHSLPPQS
FROG             MPQLNPGPWFLILIFSWLVLLTFIPPKVLKHKAFNEPTTQTTEKSKPN-PWNWPWT
FRUIT FLY        MPQMAPISWL-LLFIIFSITFILF---CSINYYSYMPNSPKSNELKNINLNSMNWKW
S.cerevisiae     MPQLVPFYFMNQLTYGFLLMITLLILFSQFFLPMILRLYVSRLFISKL
A.nidulans       MPQLVPFFFVNQVVFAFIVLTVLIYAFSRYILPRLLRTYISRIYINKL
```

Fig. 1. Partial sequence of Chargerin II and Homologous Sequences of the Putative Polypeptide Encoded by Unidentified Reading Frame A6L of the Mitochondrial Gene and Related Proteins. Identical amino acids are boxed. ■, positive charge; □, negative charge.

Alternatively, based on our previous model (Higuti, 1984), the phosphate molecules bound to chargerin II may generate negative charges on the outer surface of the inner mitochondrial membrane. In our previous model, we proposed that $\Delta\mu H^+$ causes a conformational change of protein in F_0 by electrophoretic migrating of the cluster of negative charges of the protein to near the surface of the M-side of F_0 and of the cluster of positive charges of the protein to near the surface of the M-side of F_0. Thus $\Delta\mu H^+$ is transduced to energy of conformational change of the protein from the resting-form (R-form) to the energized form (E-form). Then, spontaneous reversal of the protein from the E-form to the R-form causes conformational change of ATP-binding subunit of F_1 and release of ATP formed non-energetically on its subunit into the medium (Rosen et al., 1979; Sakamoto and Tonomura, 1983).

Chargerin II purified in the present work could have such function described above. This idea may be supported by the previous finding that antibody against chargerin II inhibited ATP-Pi exchange reaction in an energy-dependent fashion (Higuti et al., 1987b). However, further studies are required to confirm this idea.

ACKNOWLEDGMENTS

We thank Drs. Susumu Tsunasawa (Institute for Protein Research, Osaka University) and Takeharu Masaki (Department of Agricultural Chemistry, Ibaraki University) for helpful discussion concerning digestion of chargerin II with lysylendopeptidase. This work was supported by grants from the Naito Foundation, The Foundation of Sanyo Broadcasting, Takarazuka Research Center of Sumitomo Chemical Co., Takeda Science Foundation and Yamada Science Foundation, and a Grant-in-Aid for Scientific Research (Nos. 60480502 and 62580124) from the Ministry of Education, Science and Culture of Japan.

REFERENCES

Amzel, L.M. and Pedersen, P.L. (1983) Ann. Rev. Biochem. 62, 801-824.

Anderson, S., Bankier, A.T., Barrell, B.G., de Bruijn, M.H.L., Coulson, A.R., Drouin, J., Eperon, I.C., Nierlich, D.P., Roe, B.A., Sanger, F., Schreier, P.H., Smith, A.J.H., Staden, R. and Young, I.G. (1981) Nature 290, 457-465.

Anderson, S., de Bruijn, M.H.L., Coulson, A.R., Eperon, I.C., Sanger, F. and Young, I.G. (1982) J. Mol. Biol. 156, 683-717.

Bibb, M.J., Etten, A.V., Wright, C.T., Walberg, M.W. and Clayton, D.A. (1981) Cell 26, 167-180.

Blondin, G.A. (1979) Biochem. Biophys. Res. Commun. 87, 1087-1094.

Burnette, W.N. (1981) Anal. Biochem. 112, 195-203.

Chomyn, A., Mariottini, P., Gonzalez-Cadavid, N., Attardi, G., Strong, D. D., Trovato, D., Riley, M. and Doolittle, R. F. (1983) Proc. Natl. Acad. Sci. USA 80, 5535-5539.

Fearnley, I.M. and Walker, J.E. (1986) EMBO J. 5, 2003-2008.

Grisi, E., Brown, T.A., Waring, R.B., Scazzocchio, C. and Davies, R.W. (1982) Nucleic Acids Res. 10, 3531-3539.

Grosskopf, R. and Feldmann, H. (1981) Current Genet. 4, 151-158.

Hatefi, Y. (1985) Ann. Rev. Biochem. 54, 1015-1069.

Higuti, T. (1984) Mol. Cell. Biochem. 61, 37-61.

Higuti, T., Negama, T., Takigawa, M., Uchida, J., Yamane, T., Asai, T., Tani, I., Oeda, K., Shimizu, M., Nakamura, K. and Ohkawa, H. (1987a) J. Biol. Chem., submitted.

Higuti, T., Ohe, T., Arakaki, N. and Kotera, Y. (1981) J. Biol. Chem. 256, 9855-9860.

Higuti, T., Takigawa, M., Kotera, Y., Oka, H., Uchida, J., Arakaki, R., Fujita, T. and Ogawa, T. (1985) Proc. Natl. Acad. Sci. USA 82, 1331-1335.

Higuti, T., Uchida, J., Takigawa, M., Yamane, T. and Negama, T. (1987b) Biochem. Biophys. Res. Commun. submitted.

Kagawa, Y. (1984) in *Bioenergetics* (Ernster, L., ed.) New Comprehensive Biochemistry Vol. 9, pp. 149-186, Elsevier Science Publishers, Amsterdam.

Kanazawa, H. and Futai, M. (1982) Ann. N. Y. Acad. Sci. 402, 45-64.

Lauquin, G., Pougeois, R. and Vignais, P. V. (1980) Biochemistry 19, 4620-4626.

Linnane, A.W., Lukins, H.B., Nagley, P., Marzuki, S., Hadikusumo, R.G., Jean-Francois, M.J.B., John, U.P., Ooi, B.G., Watkins, L., Willson, T.A., Wright, J. and Meltzer, S. (1985) in *Achievements and Perspectives of Mitochondrial Research: Vol. I. Bioenergetics* (Quagliariello, E., Slater, E.C., Palmieri, F., Saccone, C. and Kroon A.M., eds.) pp. 211-222, Elsevier Science Publishers, Amsterdam.

Macreadie, I.G., Choo, W.M., Novitski, C.E., Marzuki, S., Nagley, P., Linnane, A.W. and Lukins, H.B. (1982) Biochem. Int. 5, 129-136.

Macreadie, I.G., Novitski, C.E., Maxwell, R.J., John, U., Ooi, B.G., McMullen, G.L., Lukins, H.B., Linnane, A.W. and Nagley, P. (1983) Nucleic Acids Res. 11, 4435-4451.

Masaki, T., Tanabe, M., Nakamura, K. and Soejima, M. (1981) Biochim. Biophys. Acta 660, 44-50.

Mitchell, P. (1961) Nature 191, 423-427.

Pepe, G., Holtrop, M., Gadaleta, G., Kroon, A.M., Catatore, P., Gallerani, R., Benedetto, C.D., Quagliariello, C., Sbisa, E. and Saccone, C. (1983) Biochem. Int. 6, 553-563.

Rosen, G., Gresser, M., Vinkler, C. and Boyer, P.D. (1979) J. Biol. Chem. 254, 10654-10661.

Sakamoto, J. and Tonomura, Y. (1983) J. Biochem. 93, 1601-1614.

Soper, J.W. and Pedersen, P.L. (1979) Methods Enzymol. 60, 328-333.

Tommasino, M., Prezioso, G. and Palmieri, F. (1987) Biochim. Biophys. Acta 890, 39-46.

Towbin, H., Staehelin, T. and Gordon, J. (1979) Proc. Natl. Acad. Sci. USA 76, 4350-4354.

Velours, J. and Guerin, B. (1986) Biochem. Biophys. Res. Commun. 138, 78-86.

Velours, J., Esparza, M., Hoppe, J., Sebald, W. and Guerin, B. (1984) EMBO J. 3, 207-212.

Wohlrab, H. (1980) J. Biol. Chem. 255, 8170-8173.

Wolstenholme, D.R. and Clary, D.O. (1985) Genetics 109, 725-744.

Wong, J.F.H., Ma, D.P., Wilson, R.K. and Roe, B.A. (1983) Nucleic Acids Res. 11, 4977-4995.

CONTRIBUTORS

INDEX

Protease, 80, 81, 88, 89, 98-100, 131, 165, 171, 260
Protein import, 74, 76, 80-82, 85-90, 95-100
Proteolipid subunit, 20, 21, 42, 51-59, 64, 90-93, 95-102, 118-122, 164
Proton gate, 184
Pseudogene, 38, 39, 41

Rapid flow-quench assay, 184
Rat, 136, 152, 165, 195, 197, 199, 200, 239, 240, 273-275, 277
Reconstitution, 3, 11, 14, 45, 48, 49, 100-136, 151-156, 165, 166, 197, 198, 212-216, 229-236
RFLP, 69, 73, 77, 81
Rhodospirillum rubrum, 229
Rotational Mechanism, 227, 228

Saccharomyces cerevisiae, 59, 85, 95-100, 117, 121, 124
Single site catalysis, 252, 253, 255
Site-directed mutagenesis, 3, 4, 11, 65

Spinach, 20-24, 29-32, 73, 74, 148, 219, 267, 268
Stabilizing factor, 173-179
Subunit 6, 21, 52, 59-64, 99, 102, 122, 124, 164-166, 201, 202
Subunit *a*, 6, 21, 24, 201, 202
Subunit *b*, 21, 113, 121, 165
Subunit *c*, 20, 21, 31, 113, 200-202
Subunit III, 20, 21, 60
Suppressor mutation, 64, 65, 85, 91-93
Sweet potato, 36, 37
Synaptic vesicles, 129, 136-139
Synechococcus, 5, 22,

TF_1, 3-7, 269
TF_0, 3, 5
Thermophilic ATP synthase, 3-7, 108, 269
Tobacco, 28, 32, 37, 69, 143
Transcripts, 22, 28-33, 39-42, 73, 75, 78-90, 99-102
Two dimensional crystals, 108, 110, 111

Unc operon, 9-14, 32, 46-48, 59, 120
URFA6L, 274-277